PENGUIN BOOKS

DIGITAL GOLD

'Lively and thorough . . . a vivid guide to the characters who met online and built Bitcoin' John Gapper, *Financial Times*

'Totally awesome' Justin Fox, *Bloomberg*

'Finally, the book so many of us have been waiting for' Adam Davidson, co-founder of NPR's Planet Money

'A very human story . . . highly entertaining' Larry Summers, former Secretary of the Treasury of the United States of America

'Paints a vivid portrait of the economics and technology of Bitcoin as well as the people behind it' Susan Athey, The Economics of Technology Professor, Stanford Graduate School of Business and Winner of the John Bates Clark Medal in Economics

'If you want to understand the future of money, read this book' Joshua Davis, author of *Spare Parts*

'A comprehensive account of the early days of what looks destined to become a new way of approaching the very nature of money' Dominic Lenton, *Engineering & Technology*

'This excellent work is the book on Bitcoin you've been waiting for' Tyler Cowen, *Marginal Revolution*

'*Digital Gold* is deeply reported . . . as close as you can get to being the definitive account of the history of Bitcoin' Felix Salmon, senior editor, *Fusion*

'An engrossing look at a system creatively designed to bring money into the twenty-first century' *Library Journal*

'*Digital Gold* is as strong a narrative achievement as a reporting one' Chris Wilson, *Bookforum*

'Bitcoin may be inherently speculative, but *Digital Gold* is a sound investment' Edward Chancellor, *Reuters*

Nathaniel Popper is a reporter at *The New York Times*. Before joining *The Times*, he worked at the *Los Angeles Times* and *The Forward*. Nathaniel grew up in Pittsburgh and is a graduate of Harvard College. He lives in Brooklyn with his family.

NATHANIEL POPPER

Digital Gold

The Untold Story of Bitcoin

PENGUIN BOOKS

PENGUIN BOOKS

UK | USA | Canada | Ireland | Australia
India | New Zealand | South Africa

Penguin Books is part of the Penguin Random House group of companies
whose addresses can be found at global.penguinrandomhouse.com.

First published in the United States of America by
HarperCollins Publishers 2015
First published in Great Britain by Allen Lane 2015
Published in Penguin Books 2016

012

Set in 9.45/14.42 pt by Sabon LT Std
Printed and bound in Great Britain by Clays Ltd, Elcograf S.p.A.

A CIP catalogue record for this book is available from the British Library

ISBN: 978–0–241–18099–0

www.greenpenguin.co.uk

Penguin Random House is committed to a
sustainable future for our business, our readers
and our planet. This book is made from Forest
Stewardship Council® certified paper.

FOR MY MOM AND DAD

INTRODUCTION

●

I t was after midnight and many of the guests had already gone
to bed, leaving behind their amber-tailed tumblers of high-
end whiskey. The poker dealer who had been hired for the
occasion from a local casino had left a half hour earlier, but the
remaining players had convinced her to leave the table and cards
so that they could keep playing. The group still hovering over the
felt and chips was dwarfed by the vaulted, wood-timbered ceiling, three stories up. The large wall of windows on the far side of
the table looked out onto a long dock, bobbing on the shimmer-
ing surface of Lake Tahoe.

Sitting at one end of the table, with his back to the lake, twenty-
nine-year-old Erik Voorhees didn't look like someone who three
years earlier had been unemployed, mired in credit card debt, and
doing odd jobs to pay for an apartment in New Hampshire. Tonight Erik fitted right in with his suede oxfords and tailored jeans
and he bantered easily with the hedge fund manager sitting next to
him. His hairline was already receding, but he still had a distinct,
fresh-faced youthfulness to him. Showing his boyish dimples, Erik

joked about his poor performance at their poker game the night before, and called it a part of his "long game."

"I was setting myself up for tonight," he said with a broad toothy smile, before pushing a pile of chips into the middle of the table.

Erik could afford to sustain the losses. He'd recently sold a gambling website that was powered by the enigmatic digital money and payment network known as Bitcoin. He'd purchased the gambling site back in 2012 for about $225, rebranded it as SatoshiDice, and sold it a year later for some $11 million. He was also sitting on a stash of Bitcoins that he'd begun acquiring a few years earlier when each Bitcoin was valued at just a few dollars. A Bitcoin was now worth around $500, sending his holdings into the millions. Initially snubbed by investors and serious business folk, Erik was now attracting a lot of high-powered interest. He had been invited to Lake Tahoe by the hedge fund manager sitting next to him at the poker table, Dan Morehead, who had wanted to pick the brains of those who had already struck it rich in the Bitcoin gold rush.

For Voorhees, like many of the other men at Morehead's house, the impulse that had propelled him into this gold rush had both everything and nothing to do with getting rich. Soon after he first learned about the technology from a Facebook post, Erik predicted that the value of every Bitcoin would grow astronomically. But this growth, he had long believed, would be a consequence of the multi-layered Bitcoin computer code remaking many of the prevailing power structures of the world, including Wall Street banks and national governments—doing to money what the Internet had done to the postal service and the media industry. As Erik saw it, Bitcoin's growth wouldn't just make him wealthy. It would also lead to a more just and peaceful world in which governments wouldn't be able to pay for wars and individuals would have control over their own money and their own destiny.

It was not surprising that Erik, with ambitions like these, had a turbulent journey since his days of unemployment in New Hampshire. After moving to New York, he had helped convince the Winklevoss twins, Tyler and Cameron, of Facebook fame, to put almost a million dollars into a startup he helped create, called BitInstant. But that relationship ended with a knock-down, drag-out fight, after which Erik resigned from the company and moved to Panama with his girlfriend.

More recently, Erik had been spending many of his days in his office in Panama, dealing with investigators from the US Securities and Exchange Commission—one of the top financial regulatory agencies—who were questioning a deal in which he'd sold stock in one of his startups for Bitcoins. The stock had ended up providing his investors with big returns. And the regulators, by Erik's assessment, didn't seem to even understand the technology. But they were right that he had not registered his shares with regulators. The investigation, in any case, was better than the situation facing one of Erik's former partners from BitInstant, who had been arrested two months earlier, in January 2014, on charges related to money laundering.

Erik, by now, was not easily rattled. It helped that, unlike many passionate partisans, he had a sense of humor about himself and the quixotic movement he had found himself at the middle of.

"I try to remind myself that Bitcoin will probably collapse," he said. "As bullish as I am on it, I try to check myself and remind myself that new innovative things usually fail. Just as a sanity check."

But he kept going, and not just because of the money that had piled up in his bank account. It was also because of the new money that he and the other men in Lake Tahoe were helping to bring into existence—a new kind of money that he believed would change the world.

THE BITCOIN CONCEPT first came onto the scene in more modest circumstances, five years earlier, when it was posted to an obscure mailing list by a shadowy author going by the name Satoshi Nakamoto.

From the beginning, Satoshi envisioned a digital analog to old-fashioned gold: a new kind of universal money that could be owned by everyone and spent anywhere. Like gold, these new digital coins were worth only what someone was willing to pay for them—initially nothing. But the system was set up so that, like gold, Bitcoins would always be scarce—only 21 million of them would ever be released—and hard to counterfeit. As with gold, it required work to release new ones from their source, computational work in the case of Bitcoins.

Bitcoin also held certain obvious advantages over gold as a new place to store value. It didn't take a ship to move Bitcoins from London to New York—it took just a private digital key and the click of a mouse. For security, Satoshi relied on uncrackable mathematical formulas rather than armed guards.

But the comparison to gold went only so far in explaining why Bitcoin ended up attracting such attention. Each ingot of gold has always existed independent of every other ingot. Bitcoins, on the other hand, were designed to live within a cleverly constructed, decentralized network, just as all the websites in the world exist only within the decentralized network known as the Internet. Like the Internet, the Bitcoin network wasn't run by some central authority. Instead it was built and sustained by all the people who hooked their computers into it, which anyone in the world could do. With the Internet, what connected everyone together was a set of software rules, known as the Internet protocol, which governed how information moved around. Bitcoin had its own software protocol—the rules that dictated how the system worked.

The technical details of how all this worked could be mind-numbingly complicated—involving advanced math and cryptography. But from its earliest days, a small group of dedicated followers saw that at its base, Bitcoin was, very simply, a new way of creating, holding, and sending money. Bitcoins were not like dollars and euros, which are created by central banks and held and transferred by big, powerful financial institutions. This was a currency created and sustained by its users, with new money slowly distributed to the people who helped support the network.

Given that it aimed to challenge some of the most powerful institutions in our society, the Bitcoin network was, from early on, described by its followers in utopian terms. Just as the Internet took power from big media organizations and put it in the hands of bloggers and dissidents, Bitcoin held out the promise of taking power from banks and governments and giving it to the people using the money.

This was all rather high-minded stuff and it attracted plenty of derision—most ordinary folks imagined it falling somewhere on the spectrum between Tamagotchi pet and Ponzi scheme, when they heard about it at all.

But Bitcoin had the good fortune of entering the world at a utopian moment, in the wake of a financial crisis that had exposed many of the shortcomings of our existing financial and political system, creating a desire for alternatives. The Tea Party, Occupy Wall Street, and WikiLeaks—among others—had divergent goals, but they were united in their desire to take power back from the privileged elite and give it to individuals. Bitcoin provided an apparent technological solution to these desires. The degree to which Bitcoin spoke to its followers was apparent from the variety of people who left their old lives behind to chase the promise of this technology—aficionados like Erik Voorhees and many of his new friends. It didn't hurt that if Bitcoin worked, it would make the early users fabulously wealthy. As Erik liked to say, "It's the first thing I know where you can both get rich and change the world."

Given the opportunity to make money, Bitcoin was not only attracting disaffected revolutionaries. Erik's host, Dan Morehead, had gone to Princeton and worked at Goldman Sachs before starting his own hedge fund. Morehead was a leading figure among the moneyed interests who had recently been pumping tens of millions of dollars into the Bitcoin ecosystem, hoping for big returns. In Silicon Valley, investors and entrepreneurs were clamoring to find ways to use Bitcoin to improve on existing payment systems like PayPal, Visa, and Western Union and to steal Wall Street's business.

Even people who had little sympathy for Occupy Wall Street or the Tea Party could understand the benefits of a more universal money that doesn't have to be exchanged at every border; the advantages of a digital payment method that doesn't require you to hand over your identifying information each time you use it; the fairness of a currency that even the poorest people in the world can keep in a digital account without paying hefty fees, rather than relying only on cash; and the convenience of a payment system that makes it possible for online services to charge a penny or a dime—to view a single news article or skip an ad—skirting the current limits imposed by the 20- or 30-cent minimum charge for a credit card transaction.

In the end, though, many of the people interested in more practical applications of Bitcoin still ended up talking about the technology in revolutionary terms: as an opportunity to make money by disrupting the existing status quo. At the dinner a few hours before the late-night poker game, Morehead had joked about the fact that, at the time, all the Bitcoins in the world were worth about the same amount as the company Urban Outfitters, the purveyor of ripped jeans and dorm room decorations—around $5 billion.

"That's just pretty wild, right?" Morehead said. "I think when they dig up our society, all Planet of Apes–style, in a couple of centuries, Bitcoin is probably going to have had a greater impact on the world than Urban Outfitters. We're still in early days."

Many bankers, economists, and government officials dismissed the Bitcoin fanatics as naive promoters of a speculative frenzy not unlike the Dutch tulip mania four centuries earlier. On several occasions, the Bitcoin story bore out the warnings of the critics, illustrating the dangers involved in moving toward a more digitized world with no central authority. Just a few weeks before Morehead's gathering, the largest Bitcoin company in the world, the exchange known as Mt. Gox, announced that it had lost the equivalent of about $400 million worth of its users' Bitcoins and was going out of business—the latest of many such scandals to hit Bitcoin users.

But none of the crises managed to destroy the enthusiasm of the Bitcoin believers, and the number of users kept growing through thick and thin. At the time of Morehead's gathering, more than 5 million Bitcoin wallets had been opened up on various websites, most of them outside the United States. The people at Morehead's house represented the wide variety of characters who had been drawn in: they included a former Wal-Mart executive who had flown in from China, a recent college graduate from Slovenia, a banker from London, and two old fraternity brothers from Georgia Tech. Some were motivated by their skepticism toward the government, others by their hatred of the big banks, and yet others by more intimate, personal experiences. The Chinese Wal-Mart executive, for instance, had grown up with grandparents who escaped the communist revolution with only the wealth they had stored in gold. Bitcoin seemed to him like a much more easily transportable alternative in an uncertain world.

It was these people, in different places with different motivations, who had built Bitcoin and were continuing to do so, and who are the subject of this story. The creator of Bitcoin, Satoshi, disappeared back in 2011, leaving behind open source software that the users of Bitcoin could update and improve. Five years later, it was estimated that only 15 percent of the basic Bitcoin computer code was the same as what Satoshi had written. Beyond the work on the software, Bitcoin, like all money, was always only as useful and powerful as the number of people using it. Each new person who joined in made it that much more likely to survive.

This, then, is not a normal startup story, about a lone genius molding the world in his image and making gobs of money. It is, instead, a tale of a group invention that tapped into many of the prevailing currents of our time: the anger at the government and Wall Street; the battles between Silicon Valley and the financial industry; and the hopes we have placed in technology to save us from our own human frailty, as well as the fear that the power of technology can generate. Each of the people discussed in this book had his or her own reason for chasing this new idea, but all their lives have been shaped by the ambitions, greed, idealism, and human frailty that have elevated Bitcoin from an obscure academic paper to a billion-dollar industry.

For some participants, the outcome has been the type of wealth on display at Morehead's house, where the stone entranceway is decorated with Morehead's personal heraldic crest. For others, it has ended in poverty and even prison. Bitcoin itself is always one big hack away from total failure. But even if it does collapse, it has already provided one of the most fascinating tests of how money works, who benefits from it, and how it might be improved. It is unlikely to replace the dollar in five years, but it provides a glimpse of where we might be when the government inevitably stops printing the faces of dead presidents on expensive paper.

The morning after the big poker game, as the guests were packing up to go, Voorhees sat at the end of the pier behind Morehead's house, which was sitting high above the water after a winter with little snowfall. The joy he had shown at the poker table the night before was gone. He had a look of chagrin on his face as he talked about his recent decision to resign as the CEO of the Bitcoin startup he had been running in Panama. His position with the company had prevented him from speaking about the revolutionary potential of Bitcoin, for fear that it could hurt his company.

"My passion is not running a business, it is building the Bitcoin world," he explained.

On top of that, his girlfriend had grown tired of living in Panama and Erik was missing his family back in the United States. In a few weeks he was planning to move back to Colorado, where he grew up. Because of Bitcoin, though, he would be going home a very different person from what he was when he left. It was a situation that many of his fellow Bitcoiners could sympathize with.

PART ONE

CHAPTER 1

●

January 10, 2009

t was a Saturday. It was his son's birthday. The Santa Barbara weather was beautiful. And his sister-in-law was in from France. But Hal Finney needed to be at his computer. This was a day he had been anticipating for months and, in some sense, for decades.

Hal didn't even try to explain to his wife, Fran, what was occupying him. She was a physical therapist and rarely understood his computer work. But with this one, where would he even begin? *Honey, I'm going to try to make a new kind of money.*

That, in essence, was his intention when, after a long morning run, he sat down in his modest home office: a corner of his living room with an old sectional desk, taken up primarily by four computer screens of different shape and make, all wired to the separate computers he used for work and personal pursuits. Any space that wasn't occupied by computer equipment was covered in a jumble of papers, exercise books, and old programming manuals. It wasn't much to look at. But sitting there, Hal could see his patio on the other side of his living room, bathed in California sun, even in the middle of January. On the carpet to his left lay Arky, his faithful Rhodesian ridgeback, named after a star in the constellation Boötes. This was where he felt at home, and where he had done much of his most creative work as a programmer.

He fired up his hulking IBM ThinkCentre, settled in, and clicked on the website he'd gotten in an e-mail the previous day while he was at work: www.bitcoin.org.

Bitcoin had first crossed his screen a few months earlier, in a message sent to one of the many mailing lists he subscribed to. The back-and-forth was usually between the familiar personalities he'd been talking to for years who inhabited the relatively specialized corner of coding where he worked. But this particular e-mail came from an unfamiliar name—Satoshi Nakamoto—and it described what was referred to as an "e-cash" with the catchy name Bitcoin. Digital money was something Hal had experimented with for a long time, enough to make him skeptical about whether it could ever work. But something jumped out in this e-mail. Satoshi promised a kind of cash that wouldn't need a bank or any other third party to manage it. It was a system that could live entirely in the collective computing memory of the people who used it. Hal was particularly drawn to Satoshi's claim that users could own and trade Bitcoins without providing identifying information to any central authorities. Hal had spent most of his professional life working on programs that allowed people to elude the ever-watchful gaze of the government.

After reading the nine-page description, contained in what looked like an academic paper, Hal responded enthusiastically:

"When Wikipedia started I never thought it would work, but it has proven to be a great success for some of the same reasons," he wrote to the group.

In the face of skepticism from others on the e-mail list, Hal had urged Satoshi to write up some actual code for the system he had described. A few months later, on this Saturday in January,

Hal downloaded Satoshi's code from the Bitcoin website. A simple .exe file installed the Bitcoin program and automatically opened up a crisp-looking window on his computer desktop.

When the program opened for the first time it automatically generated a list of Bitcoin addresses that would be Hal's account numbers in the system and the password, or private key, that gave him access to each address. Beyond that, the program had only a few functions. The main one, "Send Coins," didn't seem like much of an option for Hal given that he didn't have any coins to send. But before he could poke around further the program crashed.

It didn't deter Hal. After looking at his computer logs, he wrote to Satoshi to explain what had happened when his computer had tried to link up with other computers on the network. Apart from Hal, the log showed that there were only two other computers on the network and both of those were from a single IP address, presumably Satoshi's, tied to an Internet provider in California.

Within an hour, Satoshi had written back, expressing disappointment with the failure. He said he'd been testing it heavily and never encountered any trouble. But he told Hal that he had trimmed down the program to make it easier to download, which must have introduced the problem.

"I guess I made the wrong decision," Satoshi wrote with palpable frustration.

Satoshi sent Hal a new version of the program, with some of the old material restored, and thanked Hal for his help. When it, too, crashed, Hal kept at it. He finally got it running using a program that operated outside Microsoft Windows. Once it was up, he clicked on the most exciting-sounding function in the dropdown menu: "Generate Coins." When he did this, the processor in his computer audibly clicked into gear at a high clip.

With everything running, Hal could take a break and attend to his familial duties, including a family dinner at a nearby Chinese restaurant and a small birthday party for his son. The instructions Satoshi had included with the software said that actually generating coins could take "days or months, depending on the speed of your computer and the competition on the network."

Hal dashed off a quick note telling Satoshi that everything was working: "I have to go out but I'll leave this version running for a while."

Hal had already read enough to understand the basic work his computer was doing. Once the Bitcoin program was running, it logged into a designated chat channel to find other computers running the software—basically just Satoshi's computers at this point. All the computers were trying to capture new Bitcoins, which were released into the system in bundles of fifty coins. Each new block of Bitcoin was assigned to the address of one user who linked into the network and won a race of sorts to solve a computational puzzle. When a computer won one round of the race and captured new coins, all the other machines on the network updated their shared record of the number of Bitcoins owned by that computer's Bitcoin address. Then the computers on the network would automatically begin racing to solve a new problem to unlock the next batch of fifty coins.

When Hal returned to his computer in the evening, he immediately saw that it had made him 50 Bitcoins, now recorded next to one of his Bitcoin addresses and also recorded on the public ledger that kept track of all Bitcoins. These, the seventy-eighth block of coins generated, were among the first 4,000 Bitcoins to make it into the real world. At the time they were worth exactly nothing, but that didn't dampen Hal's enthusiasm. In a congratulatory e-mail to Satoshi that he sent to the entire mailing list, he allowed himself a flight of fancy.

"Imagine that Bitcoin is successful and becomes the dominant payment system in use throughout the world," he wrote. "Then the total value of the currency should be equal to the total value of all the wealth in the world."

By his own calculations, that would make each Bitcoin worth some $10 million.

"Even if the odds of Bitcoin succeeding to this degree are slim, are they really 100 million to one against? Something to think about," he wrote before signing off.

HAL FINNEY HAD long been preoccupied by how, in look and texture, the future would be different from the present.

One of four children of an itinerant petroleum engineer, Hal had worked his way through the classics of science fiction, but he also read calculus books for fun and eventually attended the California Institute of Technology. He never backed down from an intellectual challenge. During his freshman year he took a course on gravitational field theory that was designed for graduate students.

But he wasn't a typical nerd. A big, athletic guy who loved to ski in the California mountains, he had none of the social awkwardness common among Cal Tech students. This active spirit carried over into his intellectual pursuits. When he read the novels of Larry Niven, which discussed the possibility of cryogenically freezing humans and later bringing them back to life, Hal didn't just ponder the potential in his dorm room. He located a foundation dedicated to making this process a reality and signed up to receive the Alcor Life Extension Foundation's magazine. Eventually he would pay to have his and his family's bodies put into Alcor's frozen vaults near Los Angeles.

The advent of the Internet had been a boon for Hal, allowing him to connect with other people in far-flung places who were thinking about similarly obscure but radical ideas. Even before the invention of the first web browser, Hal joined some of the earliest online communities, with names like the Cypherpunks and Extropians, where he jumped into debates about how new technology could be harnessed to shape the future they all were dreaming up.

Few questions obsessed these groups more than the matter of how technology would alter the balance of power between corporations and governments on one hand and individuals on the other. Technology clearly gave individuals unprecedented new powers. The nascent Internet allowed these people to communicate with kindred spirits and spread their ideas in ways that had previously been impossible. But there was constant discussion of how the creeping digitization of life also gave governments and companies more command over perhaps the most valuable and dangerous commodity in the information age: information.

In the days before computers, governments certainly kept records about their citizens, but most people lived in ways that made it impossible to glean much information about them. In the 1990s, though—long before the National Security Agency was discovered to be snooping on the cell phones of ordinary citizens and Facebook's privacy policies became a matter for national debate—the Cypherpunks saw that the digitization of life made it much easier for the authorities to harvest data about citizens, making the data vulnerable to capture by nefarious actors. The Cypherpunks became consumed by the question of how people could protect their personal information and maintain their privacy. The Cypherpunk Manifesto, delivered to the mailing list in 1993 by the Berkeley mathematician Eric Hughes, began: "Privacy is necessary for an open society in the electronic age."

This line of thinking was, in part, an outgrowth of the libertarian politics that had become popular in California in the 1970s and 1980s. Suspicion regarding government had a natural appeal for programmers like Hal, who were at work creating a new world through code, without needing to rely on anyone else. Hal had imbibed these ideas at Cal Tech and in his reading of the novels of Ayn Rand. But the issue of privacy in the Internet age had an appeal beyond libertarian circles, among human rights activists and other protest movements.

None of the Cypherpunks saw a solution to the problem in running away from technology. Instead, Hal and the others aimed to find answers in technology and particularly in the science of encrypting information. Encryption technologies had historically been a privilege largely reserved for only the most powerful institutions. Private individuals could try to encode their communications, but governments and armed forces almost always had the power to crack such codes. In the 1970s and 1980s, though, mathematicians at Stanford and MIT made a series of breakthroughs that made it possible, for the first time, for ordinary people to encrypt, or scramble, messages in a way that could be decrypted only by the intended recipient and not cracked even by the most powerful supercomputers.

Every user of the new technology, known as public-key cryptography, would receive a public key—a unique jumble of letters and numbers that serves as a sort of address that could be distributed freely—and a corresponding private key, which is supposed to be known only by the user. The two keys are related, mathematically, in a way that ensures that only the user—let's call her Alice, as cryptographers often did—with her private key, can unlock messages sent to her public key, and only she can sign off on messages associated with her public key. The unique relationship between each public and private key was determined by complicated math equations that were constructed so cleverly that no one with a particular public key would ever be able to work backward to figure out the corresponding private key—not even the most powerful supercomputer. This whole setup would later play a central role in the Bitcoin software.

Hal was introduced to the potential of public-key cryptography in 1991 by the pathbreaking cryptographer David Chaum, who had been experimenting with ways to use public-key cryptography to protect individual privacy.

"It seemed so obvious to me," Hal told the other Cypherpunks of his first encounter with Chaum's writing. "Here we are faced with the problems of loss of privacy, creeping computerization, massive databases, more centralization—and Chaum offers a completely different direction to go in, one which puts power into the hands of individuals rather than governments and corporations."

As usual, when Hal found something exciting, he didn't just passively read up on it. On nights and weekends, after his job as a software developer, he began helping with a volunteer project, referred to as Pretty Good Privacy, or PGP, which allowed people to send each other messages that could be encrypted using public-key cryptography. The founder of the project, Phil Zimmerman, was an antinuclear activist who wanted to give dissidents a way to communicate outside the purview of governments. Before long, Zimmerman brought Hal on as the first employee at PGP.

Idealistic projects like PGP generally had a small audience. But the potential import of the technology became apparent when federal prosecutors launched a criminal investigation into PGP and Zimmerman. The government categorized encryption technology, such as PGP, as weapon-grade munitions, and this designation made it illegal to export. While the case was eventually dropped, Hal had to lie low with his own involvement in PGP for years and could never take credit for some of his important contributions to the project.

•

THE EXTROPIANS AND Cypherpunks were working on several different experiments that could help empower individuals against traditional sources of authority. But money was, from the beginning, at the center of their efforts to reimagine the future.

Money is to any market economy what water, fire, or blood is to the human ecosystem—a basic substance needed for everything else to work. For programmers, existing currencies, which were valid only within particular national borders and subject to technologically incompetent banks, seemed unnecessarily constrained. The science fiction that Hal and others had grown up on almost always featured some kind of universal money that could span galaxies—in *Star Wars* it was the galactic credit standard; in the *Night's Dawn* trilogy it was Jovian credit.

Beyond these more fanciful ambitions, the existing financial system was viewed by the Cypherpunks as one of the biggest threats to individual privacy. Few types of information reveal as much about a person like Alice, the cryptographers' favorite, as her financial transactions. If snoopers get access to her credit card statements they can follow her movements over the course of a day. It's no accident that financial records are one of the primary ways that fugitives are tracked down. Eric Hughes's Cypherpunk Manifesto had dwelled on this problem at great length: "When my identity is revealed by the underlying mechanism of the transaction, I have no privacy. I cannot here selectively reveal myself; I must always reveal myself," Hughes wrote.

"Privacy in an open society requires anonymous transaction systems," he added.

Cold, hard cash had long provided an anonymous way of making payments, but this cash did not make the transition over to the digital realm. As soon as money became digital, some third party, such as a bank, was always involved and therefore able to trace the transaction. What Hal, Chaum, and the Cypherpunks wanted was a cash for the digital age that could be secure and uncounterfeitable without sacrificing the privacy of its users. The same year as Hughes's manifesto, Hal wrote an e-mail to the group imagining a kind of digital cash for which "no records are kept of where I spend my money. All the bank knows is how much I have withdrawn each month."

A month later, Hal even came up with a cheeky moniker for it: "I thought of a new name today for digital cash: CRASH, taken from CRypto cASH."

Chaum himself had already come up with his own version of this by the time the Cypherpunks got interested. Working out of an institute in Amsterdam, he had created DigiCash, an online money that could be spent anywhere in the world without requiring users to hand over any personal information. The system harnessed public-key cryptography to allow for what Chaum called blind digital signatures, which allowed people to sign off on transactions without providing any identifying information. When Mark Twain Bank in the United States began experimenting with DigiCash, Hal signed up for an account.

But Chaum's effort would rub Hal and others the wrong way. With DigiCash, a central organization, namely Chaum's company, needed to confirm every digital signature. This meant that a certain degree of trust needed to be placed in that central organization not to tinker with balances or go out of business. Indeed, when Chaum's company went bankrupt in 1998, DigiCash went down with it. These concerns pushed Hal and others to work toward a digital cash that wouldn't rely on any central institution. The problem, of course, was that someone needed to check that people weren't simply copying and pasting their digital money and

DIGITAL GOLD 13

spending it twice. Some of the Cypherpunks simply gave up on the project, but Hal wasn't one to fold so easily.

Ironically for a person so eager to create new money, Hal's interest wasn't primarily financial. The programs he was writing, like PGP, were explicitly designed to be available to anyone, free. His political distrust of government, meanwhile, was not driven by selfish resentment about paying taxes. During the 1990s Hal would calculate the maximum bill for his tax bracket and send in a check for that amount, so as to avoid the hassle of actually filling out a return. He bought his modest home on the outskirts of Santa Barbara and stuck with it over the years. He didn't seem to mind that he had to work out of his living room or that the blue recliners in front of his desk were wearing thin. Instead of being motivated by self-interest, his work seemed driven by an intellectual curiosity that bubbled over in each e-mail he wrote, and by his sense of what he thought other people deserved.

"The work we are doing here, broadly speaking, is dedicated to this goal of making Big Brother obsolete. It's important work," Hal would write to his fellow travelers. "If things work out well, we may be able to look back and see that it was the most important work we have ever done."

CHAPTER 2

●

1997

T he notion of creating a new kind of money would seem, to many, a rather odd and even pointless endeavor. To most modern people, money is always and everywhere bills and coins issued by countries. The right to mint money is one of the defining powers of a nation, even one as small as the Vatican City or Micronesia.

But that is actually a relatively recent state of affairs. Until the Civil War, a majority of the money in circulation in the United States was issued by private banks, creating a crazy patchwork of competing bills that could become worth nothing if the issuing bank went down. Many countries at that time relied on circulating coins from other countries.

This was the continuation of a much longer state of affairs in which humans engaged in a seemingly ceaseless effort to find better forms of money, trying out gold, shells, stone disks, and mulberry bark along the way.

The search for a better form of money has always been about finding a more trustworthy and uniform way of valuing the things around us—a single metric that allows a reliable comparison between the value of a block of wood, an hour of carpentry work, and a painting of a forest. As sociologist Nigel Dodd put it, good money is "able to convert qualitative differences between things into quantitative differences that enable them to be exchanged."

The money imagined by the Cypherpunks looked to take the standardizing character of money to its logical extreme, allowing for a universal money that could be spent anywhere, unlike the constrained national currencies we currently carry around and exchange at each border.

In their efforts to design a new currency, the Cypherpunks were mindful of the characteristics usually found in successful coinage. Good money has generally been durable (imagine a dollar bill printed on tissue paper), portable (imagine a quarter that weighed twenty pounds), divisible (imagine if we had only hundred-dollar bills and no coins), uniform (imagine if all dollar bills looked different), and scarce (imagine bills that could be copied by anyone).

But beyond all these qualities, money always required something much less tangible and that was the faith of the people using it. If a farmer is going to accept a dollar bill for his hard-earned crops, he has to believe that the dollar, even if it is only a green piece of paper, will be worth something in the future. The essential quality of successful money, through time, was not who issued it—or even how portable or durable it was—but rather the number of people willing to use it.

In the twentieth century, the dollar served as the global currency in no small part because most people in the world believed that the United States and its financial system had a better chance of surviving than almost anything else. That explains why people sold their local currency to keep their savings in dollars.

Money's relationship to faith has long turned the individuals who are able to create and protect money into quasi-religious figures. The word *money* comes from the Roman god Juno Moneta, in whose temple coins were minted. In the United States, the governors of the central bank, the Federal Reserve, who are tasked with overseeing the money supply, are treated like oracles of sorts; their pronouncements are scrutinized like the goat entrails of olden days. Fed officials are endowed with a level of power and independence given to almost no other government leaders, and the task of protecting the nation's currency is entrusted to a specially created agency, the Secret Service, that was only later given the additional responsibility of protecting the life of the president.

Perhaps the most famous, if flawed, oracle of the Federal Reserve, former chairman Alan Greenspan, knew that money was something that not only central bankers could create. In a speech in 1996, just as the Cypherpunks were pushing forward with their experiments, Greenspan said that he imagined that the technological revolution could bring back the potential for private money and that it might actually be a good thing:

"We could envisage proposals in the near future for issuers of electronic payment obligations, such as stored-value cards or 'digital cash,' to set up specialized issuing corporations with strong balance sheets and public credit ratings."

IN THE YEARS right after Greenspan's speech, there was a flurry of activity in the Cypherpunk world. In 1997 a British researcher named Adam Back released on the Cypherpunk mailing list his plan for something he called hashcash, which solved one of the most basic problems holding back the digital-cash project: the seeming impossibility of creating any sort of digital file that can't be endlessly copied.

To solve this problem, Back had a clever idea, which would later be an important building block for the Bitcoin software. Back's concept made creative use of one of the central cogs of public-key cryptography: cryptographic hash functions. These are math equations that are easy to solve but hard to reverse-engineer, just as it is relatively easy to multiply 2,903 and 3,571 using a piece of paper and pencil, but much, much harder to figure out what two numbers can be multiplied together to get 10,366,613. With hashcash, computers essentially had to figure out which two numbers can be multiplied together to get 10,366,613, though the problems for hashcash were significantly harder than that. So hard, in fact, that all a computer could do was try out lots of different guesses with the aim of eventually finding the right answer. When a computer found the right answer, it would earn hashcash.

The creation of hashcash through this method was useful in the context of digital money because it ensured that hashcash would be scarce—a characteristic of most good money but not of digital files, which are generally easily duplicated. A computer had to perform lots of work to create each new unit of hashcash, earning the process the name "proof-of-work"—something that would later be a central innovation underpinning Bitcoin. The main problem with Back's system, as a type of digital money, was that each hashcash unit could be used only once and everyone in the system needed to create new units whenever they wanted to use any. Another problem was that a person with unlimited computing power could produce more and more hashcash and reduce the overall value of each unit.

A year after Back released his program, two different members of the Cypherpunk list came up with systems that solved some of hashcash's shortcoming, creating digital tokens that required a proof-of-work, but that could also be reused. One of these, a concept called bit gold, was invented by Nick Szabo, a security expert and Cypherpunk who circulated his idea to close collaborators like Hal Finney in 1998, but never actually put it into practice.

Another, known as b-money, came from an American named Wei Dai. Hal created his own variant, with a decidedly less sexy name: reusable proofs of work, or RPOWs.

The conversation around these ideas on the Cypherpunk list and among related groups sometimes resembled the bickering of rivalrous brothers trying to one-up each other. Szabo would snipe at other proposals, saying that they all relied too much on specialized computer hardware instead of software. But these men—and they were all men—also built up deep respect for each other. And even as their experiments failed, their ambitions grew beyond just anonymous money. Among other things, Back, Szabo, and Finney sought to overcome the costs and frustrations of the current financial system in which banks charged fees with every transaction and made it difficult to move money over international borders.

"What we want is fully anonymous, ultra low transaction cost, transferable units of exchange. If we get that going (and obviously there are some people trying DigiCash, and a couple of others), the banks will become the obsolete dinosaurs they deserve to become," Back told the Cypherpunk list soon after releasing hashcash.

The Cypherpunk seekers were given a platonic ideal to shoot for when science fiction writer Neal Stephenson published his book *Cryptonomicon* in 1999. The novel, which became legendary in hacker circles, imagined a subterranean world that was fueled by a kind of digital gold that allowed people to keep their identities private. The novel included lengthy descriptions of the cryptography that made it all possible.

But the experiments that the Cypherpunks were doing in the real world continued to hit practical hurdles. No one could figure out a way to create money without relying on a central institution that was vulnerable to failure or government oversight. The experiments also suffered from a more fundamental difficulty, which was the issue of getting people to use and value these new digital tokens. By the time Satoshi Nakamoto came onto the scene, history had made many of Bitcoin's most likely fans very jaded. The goal of creating digital money seemed as much of a dream as turning coal into diamonds.

IN AUGUST 2008 Satoshi emerged out of the mists in an e-mail sent to the creator of hashcash, Adam Back, asking him to look at a short paper describing something called Bitcoin. Back hadn't heard of it or Satoshi, and didn't spend much time on the e-mail, other than to point Satoshi to other Cypherpunk experiments that he might have missed.

Six weeks later, on Halloween, Satoshi sent a more fleshed-out proposal to a specialized, and heavily academic, mailing list focused on cryptography—one of the main successors to the Cypherpunk list, which was defunct. As was typical in this community, Satoshi gave no information about his own identity and background, and no one asked. What mattered was the idea, not the person. In careful, dry language, Satoshi opened with a bold claim to have solved many of the problems that had dogged the long search for the holy grail of universal money.

"I've been working on a new electronic cash system that's fully peer-to-peer, with no trusted third party," the e-mail began.

The nine-page PDF attached to the e-mail made it clear that Satoshi was deeply versed in all the previous efforts to create a self-sustaining digital money. Satoshi's paper cited Back and Wei Dai, as well as several obscure journals of cryptography. But Satoshi put all these earlier innovations together to create a system that was quite unlike anything that had come before it.

Rather than relying on a central bank or company to issue and keep track of the money—as the existing financial system and Chaum's DigiCash did—this system was set up so that every

Bitcoin transaction, and the holdings of every user, would be tracked and recorded by the computers of all the people using the digital money, on a communally maintained database that would come to be known as the blockchain.

The process by which this all happened had many layers, and it would take even experts months to understand how they all worked together. But the basic elements of the system can be sketched out in rough terms, and were in Satoshi's paper, which would become known as the Bitcoin white paper.

According to the paper, each user of the system could have one or more public Bitcoin addresses—sort of like bank account numbers—and a private key for each address. The coins attached to a given address could be spent only by a person with the private key corresponding to the address. The private key was slightly different from a traditional password, which has to be kept by some central authority to check that the user is entering the correct password. In Bitcoin, Satoshi harnessed the wonders of public-key cryptography to make it possible for a user—let's call her Alice again—to sign off on a transaction, and prove she has the private key, without anyone else ever needing to see or know her private key.*

Once Alice signed off on a transaction with her private key she would broadcast it out to all the other computers on the Bitcoin network. Those computers would check that Alice had the coins she was trying to spend. They could do this by consulting the public record of all Bitcoin transactions, which computers on the network kept a copy of. Once the computers confirmed that Alice's address did indeed have the money she was trying to spend, the information about Alice's transaction was recorded in a list of all recent transactions, referred to as a block, on the blockchain.

---

* For more detail on how this and other basic elements of how the Bitcoin network worked, see the Technical Appendix on page 357.

The exact method used to add blocks to the blockchain was perhaps the most complicated part of the system. At the simplest level, it involved a sort of computational race between all computers on the network, modeled after the contest that Adam Back had invented for hashcash. The computer that won the race was responsible for inscribing the most recent block of transactions onto the blockchain. Equally important, the winner also received a bundle of new Bitcoins—50 Bitcoins when the network actually started operating. This was, indeed, the only way new Bitcoins could be brought into the world. The reward of new coins helped encourage Bitcoin users to set their computers to partake in the communal work of recording transactions.

If there were disagreements about which computer won the lottery, the record of transactions that had already been adopted by the most computers on the network would prevail. If, for example, most of the computers on the network believed Alice won the latest race, but a few computers believed that Bob won the race, the computers that used Bob's record of transactions would be ignored by other computers on the network until they joined the majority. This democratic method of decision making was valuable because it prevented a few bad computers from going rogue and assigning themselves lots of new Bitcoins; rogue elements would have to capture a majority of the computers on the network to do this.

Alterations to the Bitcoin software, which would run on the computer of every user, would also be decided by means of this democratic model. Any user could make a change to the open source Bitcoin software, but the changes would generally be effective only when a majority of the computers on the network adopted the altered version of the software. If a lone computer began running a different version of the Bitcoin software it would essentially be ignored by the other computers and would no longer be part of the Bitcoin network.

To recap, the five basic steps of the Bitcoin process were laid out as follows:

- Alice initiates a transfer of Bitcoins from her account by signing off with her private key and broadcasting the transaction to other users.
- The other users of the network make sure Alice's Bitcoin address has sufficient funds and then add Alice's transaction to a list of other recent transactions, known as a block.
- Computers take part in a computational race to have their list of transactions, or block, added to the blockchain.
- The computer that has its block added to the blockchain is also granted a bundle of new Bitcoins.
- Computers on the network start compiling a new list of unconfirmed recent transactions, trying to win the next bundle of Bitcoins.

The result of this complicated process was something that was deceptively simple but never previously possible: a financial network that could create and move money without a central authority. No bank, no credit card company, no regulators. The system was designed so that no one other than the holder of a private key could spend or take the money associated with a particular Bitcoin address. What's more, each user of the system could be confident that, at every moment in time, there would be only one public, unalterable record of what everyone in the system owned. To believe in this, the users didn't have to trust Satoshi, as the users of DigiCash had to trust David Chaum, or users of the dollar had to trust the Federal Reserve. They just had to trust their own computers running the Bitcoin software, and the code Satoshi wrote, which was open source, and therefore available for everyone to review. If the users didn't like something about the rules set down by Satoshi's software, they could change the rules. People who joined the Bitcoin network were, quite literally, both customers and owners of both the bank and the mint.

But so far, at least, all Satoshi had done was describe this grand scheme.

DESPITE ALL THE advances described in the Bitcoin paper, a week after it was posted, when Hal Finney chimed in for the first time, there were only two responses on the cryptography mailing list. Both were decidedly negative. One noted computer security expert, John Levine, said that the system would be easily overwhelmed by malicious hackers who could spread a version of the blockchain that was different from the one being used by everyone else.

"The good guys have vastly less computational firepower than the bad guys," Levine wrote on November 2. "I also have my doubts about other issues, but this one is the killer."

Levine's concern was a valid one. The Bitcoin system Satoshi described relied on computers reaching decisions by majority rule. Early on, when there were fewer computers on the network, it would be easier to become the majority and take over. But Satoshi's hope was that there wouldn't be much of an incentive to take over the system early on, when the network was small. Later on, if there was an incentive to attack the network, that would hopefully be because the network had attracted enough members to make it hard to overwhelm.

Another longtime veteran of the Cypherpunk debates, James Donald, said that "we very, very much need a system," but the way he read the paper, the database of transactions, the blockchain, would quickly become too big for users to download.

In the weeks that followed, Hal was essentially Satoshi's only defender. On the cryptography list, Hal wrote that he wasn't terribly worried about the attackers that Levine talked about. But Hal admitted that he wasn't sure how the whole thing would work in practice, and expressed a desire to see actual computer code, rather than just a conceptual description.

"This does seem to be a very promising and original idea, and I am looking forward to seeing how the concept is further developed," Hal wrote to the group.

Hal's defense of the program led Satoshi to send him an early, beta version for testing. In test runs in November and December they worked out some of the early kinks. Not long after that, in January 2009, Satoshi sent the complete code to the list. The final software made some interesting tweaks to the system described in the original paper. It determined that new coins would be assigned approximately every ten minutes, with the hash function lottery getting harder if computers were generating coins more frequently than that.

The software also mandated that the winner of each block would get fifty coins for the first four years, twenty-five coins for the next four years, and half as much again every four years until 21 million coins were released into the world, at which point new coin generation would stop.

On the first day, when Hal downloaded the software, the network was already up and running. For the next few days, not much activity was being added to the blockchain other than a computer on the network (usually belonging to Satoshi) winning fifty coins every ten minutes or so. But on Sunday evening the first transaction took place when Satoshi sent Hal ten coins to make sure that this part of the system was working smoothly. To complete the transaction, Satoshi signed off with the private key associated with the address where the coins were stored. This transaction was broadcast to the network—essentially just Hal and Satoshi at this point—and was registered in the blockchain a few minutes later when Satoshi's computers won the next round of the hash function lottery. At that point, anyone who downloaded the software would download the entire blockchain up to the point, which included a record of the ten coins that Hal had received from Satoshi, as well as the fifty coins that Hal had won on Saturday.

In the first weeks, other early adopters were slow to buy in. Satoshi was using his own computers to help power the network. Satoshi was also doing everything possible to sell the technology, responding quickly to anyone showing the slightest interest. When a programmer in Texas wrote to Satoshi late one night, expressing his own familiarity with electronic currency and cryptography, he had an answer from Satoshi the next morning.

"We definitely have similar interests!" Satoshi wrote with innocent enthusiasm, before describing the challenge that confronted Bitcoin:

> You know, I think there were a lot more people interested in the 90's, but after more than a decade of failed Trusted Third Party based systems (DigiCash, etc.), they see it as a lost cause. I hope they can make the distinction, that this is the first time I know of that we're trying a non-trust based system.

It became clear, though, that Satoshi's program on its own was just a bunch of code, sitting on a server like so many other dreams hatched by programmers. Most of those dreams die, forgotten on a hard drive somewhere. Bitcoin needed more users and defenders like Hal to survive, and there weren't many to be found. A week after the program was released, one writer on the Cryptography mailing list wrote: "No major government is likely to allow Bitcoin in its present form to operate on a large scale."

Hal acknowledged that the author could prove to be right, but came to Satoshi's defense again: "Bitcoin has a couple of things going for it: one is that it is distributed, with no single point of failure, no 'mint,' no company with officers that can be subpoenaed and arrested and shut down."

Even Hal's enthusiasm, though, appeared to flag at times. As his computer kept working at full capacity, trying to generate new coins, he began to worry about the carbon dioxide emissions caused by all the computers racing to mint coins. After his son, Jason, complained about the wear and tear it was causing to the computer, Hal turned off the Generate Coins option. Hal also had begun to fear that with a public ledger of all transactions—even if everyone was represented by a confusing-looking address—Bitcoin might not be as anonymous as he initially thought.

And then something much worse happened. Hal's speech began slurring. He became increasingly sluggish during his marathon training. Soon, all his free moments were spent visiting doctors, trying to identify the mysterious ailment. Eventually it was diagnosed as Lou Gehrig's disease, the degenerative condition that would gradually cause all his muscles to wither away inside his body. By the time he learned this, Hal was out of the Bitcoin game. He wouldn't return until his condition was much worse and Bitcoin's was much better.

CHAPTER 3

May 2009

n early May, a few months after Hal Finney's last messages, Satoshi Nakamoto received an e-mail written in stilted but precise English.

"I have a good touch on Java and C languages from school courses (I'm studying CS), but not so very much development experience yet," read the note, signed Martti Malmi.

This was clearly not the voice of a grizzled veteran of the Cypherpunk movement like Hal. But Martti displayed something more important at this point: eagerness.

"I would like to help with Bitcoin, if there's something I can do," he wrote.

Satoshi had gotten a few promising e-mails since Hal had disappeared two months earlier, but Martti was already demonstrating more commitment than the others. Before reaching out to Satoshi, Martti had written about Bitcoin on anti-state.com, a forum dedicated to the possibility of an anarchist society organized only by the market. Using the screen name Trickster, Martti gave a brief description of the Bitcoin idea and asked for thoughts:

A widespread adoption of such a system sounds like something that could have a devastating effect on the state's
ability to feed on its livestock. What do you think about
this? I'm really excited about the thought of something
practical that could truly bring us closer to freedom in our
lifetime :-) Now we just need some convincing proof that
the software and the system work securely enough to be
taken into real use.

Martti included a link to this post in his first e-mail to Satoshi,
and Satoshi quickly read it and responded.

"Your understanding of Bitcoin is spot on," Satoshi told him.

MARTTI'S ENTHUSIASM HELPED CONFIRM the shift in strategy Satoshi had made since the beginning of the year. Back when
Satoshi had first launched the software, his writings were drily
focused on the technical specifications of the programming.

But after the first few weeks, Satoshi began emphasizing the
broader ideological motivations for the software to help win over
a broader audience, and privacy was only a part of it. In a February posting on the website of the P2P Foundation, a group dedicated to decentralized, peer-to-peer technology, Satoshi led off by
talking about problems with traditional, or fiat, currencies, a term
for money generated by government decree, or fiat.

"The root problem with conventional currency is all the trust
that's required to make it work," Satoshi wrote. "The central bank
must be trusted not to debase the currency, but the history of fiat
currencies is full of breaches of that trust."

Currency debasement was not an issue the Cypherpunks had
discussed much, but Satoshi made it clear with this posting, and
not for the last time, that he had been thinking about more than

just the concerns of the Cypherpunks when designing the Bitcoin software. The issue that Satoshi referred to here—currency debasement—was, in fact, a problem with existing monetary systems that had much more potential widespread appeal, especially in the wake of the government-sponsored bank bailouts that had occurred just a few months earlier in the United States.

Throughout history, central banks have been accused of debasing their currencies by printing too much new money—or reducing the precious metal content in coins—thus making the existing money worth less. This had been a passionate political cause, in certain circles, since the end of the gold standard, the policy by which every dollar was backed by a certain quantity of gold.

The gold standard was the most popular global monetary system at the start of the twentieth century. Not only did gold link paper money to something of physical substance; the standard also served as a mechanism for imposing restraint on central banks. The Federal Reserve and other central banks could print more money only if they managed to get their hands on more gold. If they ran out of gold, no more money and no more spending.

The restriction was suspended during the Great Depression, so that central banks around the world could print more money to stimulate the economy. After World War II, the world's leading economies went back to a quasi–gold standard, with all currencies having a set value in gold—though it was no longer possible to actually turn dollars in to collect physical gold. In 1971 Richard Nixon finally decided to cut the value of the dollar loose from any anchor and end the gold standard permanently. The dollar and most other global currencies would be worth only as much as someone was willing to pay for them. Now the value of the dollar arose from the commitment of the United States government to take it for all debts and payments.

Most economists approve of the move away from the gold standard, as it allowed central banks to be more responsive to the ups and downs of the economy, putting more money into circulation when the economy grew or when people weren't spending and the economy needed a jolt. But the policy has faced impassioned criticism, particularly from antigovernment circles, where many believe that the end of the gold standard allowed central banks to print money with no restraint, hurting the long-term value of the dollar and allowing for unbridled government spending.

Until 2008, though, this was a relatively niche issue, even among libertarians. That changed during the financial crisis, after the Federal Reserve helped bail out big banks and stimulate the economy by printing lots of money. This fanned fears that the new money flooding the market would make existing money and savings worth less. Suddenly, monetary policy was a mainstream political issue and the Fed was a sort of national villain, with "END THE FED" bumper stickers becoming a common sight. The issue became one of the first criticisms of the existing financial system that gained popular appeal after the financial crisis.

When Satoshi released Bitcoin, just months after these bank bailouts, the design provided a tidy solution for people worried about a currency with no restraints. While the Federal Reserve had no formal limits on how much new money it could create, Satoshi's Bitcoin software had rules to ensure that new Bitcoins would be released only every ten minutes or so and that the process of creating new coins would stop after 21 million were out in the world.

This apparently small detail in the system carried potentially great political significance in a world worried about unlimited printing of money. What's more, the restraints on Bitcoin creation helped deal with one of the big issues that had bedeviled earlier digital moneys—the matter of how to convince users that the money would be worth something in the future. With a hard cap on the number of

Bitcoins, users could reasonably believe that Bitcoins would become harder to get over time and thus would go up in value.

These rules were all a late addition to the code and Satoshi had not played them up early on. But now that he needed to sell it to the public, this feature of Bitcoin became a big draw. Martti Malmi, the young man who wrote to Satoshi in early May, proved the wisdom of emphasizing this. Martti didn't know cryptography but as a political junkie he was immediately drawn to Bitcoin's revolutionary potential.

"There's no central bank to debase the currency with unlimited creation of new money," Martti wrote on the anti-state.com forum.

This was the first but not the last time that the Bitcoin concept's many layers, and its openness to new interpretations, would allow the project to pick up crucial new followers.

Satoshi quickly gave Martti practical suggestions for how he could help the project. The most important was the simplest: to leave his computer on with the Bitcoin program running. Five months after Bitcoin was launched, it was still not possible to trust that someone somewhere was running the Bitcoin program. When a new person tried to join, there were often no other computers or nodes to communicate with. It also meant that Satoshi's computers were still generating almost all the coins. When Martti joined in, he quickly began winning them on his laptop, which he kept running except when he needed the computing power for his video games.

As to the more complicated programming needs, Satoshi told Martti that there was "not much that's easy right now." But, Satoshi added, the Bitcoin website did need introductory material for beginners and Martti seemed like the right person for the job.

"My writing is not that great—I am a much better coder," Satoshi wrote, encouraging Martti to try his hand.

Two days later, Martti proved Satoshi right by sending a lengthy but accessible document addressing seven basic questions, ready to be posted on the Bitcoin website. Martti provided straightforward, if occasionally stilted, answers to questions like, "Is Bitcoin safe?" and "Why should I use Bitcoin?" To answer the latter, he cited the political motivations:

> Be safe from the unfair monetary policies of the monop-
> olistic central banks and the other risks of centralized
> power over a money supply. The limited inflation of the
> Bitcoin system's money supply is distributed evenly (by
> CPU power) throughout the network, not monopolized to
> a banking elite.

Satoshi liked the document so much that Martti was quickly given full credentials for the Bitcoin website, allowing him to make any improvements he wanted. Satoshi particularly encouraged Martti to help make the site look more professional and get users up to speed.

WHEN MARTTI FOUND Bitcoin in the spring of 2009, he was in his second year at the Helsinki University of Technology. If Hal Finney was the opposite of the normal tech geek, Martti lived up to type. Lanky, with birdlike features, Martti shied away from social contact. He spoke in a slow, halting voice that sounded almost as if it were computer generated. He was happiest in his room with his computer, writing code, which he had learned to do at age twelve, or hammering away at enemies in online games, while listening to heavy metal music on headphones.

Martti's reclusive, computer-centric life led him to the ideas behind Bitcoin, and ultimately to Bitcoin itself. The Internet had allowed a teenage Martti to discover and explore political ideas that were far from the Finnish social democratic consensus. The ideas of the libertarian economists he began following, which encouraged people to create their own destiny, aligned with Martti's lone-wolf approach to life, even if it ignored the incredible education that Martti had received thanks to Finland's strong government and high taxes. Who needs the state when you have talent and ideas?

During his college years, Martti had become fascinated by the rise in Scandinavia of the Pirate Party, which promoted technology over political engagement as the way to move society. Napster and other music sharing services hadn't waited for politics to reform copyright law; they forced the world to change. As Martti pondered these ideas he began wondering whether money might be the next thing vulnerable to technological disruption. After a brief spasm of random web searches, Martti had found his way to the primitive website at Bitcoin.org.

Within a few weeks of his initial exchanges with Satoshi, Martti had totally revamped the Bitcoin website. In place of Satoshi's original version, which presented complicated descriptions of the code, Martti led off with a brief, crisp description of the big ideas, aimed at drawing in anyone with similar ideological interests.

"Be safe from the unstability caused by fractional reserve banking and the bad policies of the central banks," read the newly designed site.

The onslaught of new users was slow to arrive, however. A few dozen people downloaded the Bitcoin program in June, to add to the few hundred who had downloaded it since its original release. Most had tried it once and then turned it off. But Martti kept at it. After releasing the new website, Martti turned to the software's actual underlying code. He did not know C++, the programming language that Satoshi had written Bitcoin in, so Martti began teaching himself.

Martti had time for all of this because he failed to land a summer programming job—a failure that gave Bitcoin a much-needed boost over the next months. Martti got a part-time job through a temp agency, but he would spend many of his days and nights at the university computer lab and find himself emerging at dawn. As he learned C++, Martti was going through the laborious process of compiling his own version of the code that Satoshi had written, so that he could begin making changes to it. He and Satoshi communicated regularly and fell into an easy rapport.

While Satoshi never discussed anything personal in these e-mails, he would banter with Martti about little things. In one e-mail, Satoshi pointed to a recent exchange on the Bitcoin e-mail list in which a user referred to Bitcoin as a "cryptocurrency," referring to the cryptographic functions that made it run.

"Maybe it's a word we should use when describing Bitcoin. Do you like it?" Satoshi asked.

"It sounds good," Martti replied. "A peer to peer cryptocurrency could be the slogan."

As the year went on they also worked out other details, like the Bitcoin logo, which they mocked up on their computers and sent back and forth, coming up, finally, with a B with two lines coming out of the bottom and top.

They also batted back and forth potential improvements to the software. Martti proposed making Bitcoin launch automatically when someone turned on a computer, an easy way to get more nodes on the network.

Satoshi loved it: "Now that I think about it, you've put your finger on the most important missing feature right now that would make an order of magnitude difference in the number of nodes."

Despite Martti's relative lack of programming experience, Satoshi gave him full permission to make changes to the core Bitcoin software on the server where it was stored—something that, to this point, only Satoshi could do. Starting in August, the log of changes to the software showed that Martti was now the main actor. When the next version of Bitcoin, 0.2, was released, Satoshi gave credit for most of the improvements to Martti.

But both Satoshi and Martti were struggling with how to get more people to use Bitcoin in the first place. There were other computers on the network generating coins, but the majority of coins were still captured by Satoshi's own computers. And throughout 2009 no one else was sending or receiving any Bitcoins. This was not a promising sign.

"It would help if there was something for people to use it for. We need an application to bootstrap it," Satoshi wrote to Martti in late August. "Any ideas?"

Returning to school for the fall semester, Martti worked on several fronts to address this. He was eager to set up an online forum where Bitcoin users could meet and talk. Long before Bitcoin, online forums had been where Martti had come out of his shell as a teenager, allowing him a social ease that he never had in real-life interactions. He could almost be someone else. Indeed, when Martti and Satoshi eventually set up a new Bitcoin forum, Martti gave himself the screen name that would become his alter ego in the Bitcoin world: sirius-m.

The name had a cosmic ring to it, and conveyed that this was "sirius business," Martti thought to himself. But it also had a more playful meaning for Martti, who had used the alias in a Harry Potter role-playing game at age thirteen.

The Bitcoin forum went online in the fall of 2009 and soon attracted a few regulars. One of them, who called himself New-LibertyStandard, talked about the need for a website where people could buy and sell Bitcoins for real money. Martti had been talking with Satoshi about something similar, but he was all too glad to help NewLibertyStandard. In the very first recorded transaction of Bitcoin for United States dollars, Martti sent NewLibertyStandard 5,050 Bitcoins to use for seeding the new exchange. In return, Martti got $5.02 by PayPal.

This trade raised the obvious question of how much a Bitcoin should be worth. Given that no one had ever bought or sold one, NewLibertyStandard came up with his own method for determining its value—the rough cost of electricity needed to generate a coin, calculated using NewLibertyStandard's own electricity bill. By this measure, one dollar was worth around one thousand Bitcoins for most of October and November 2009.

For Satoshi, though, more important than buying and selling Bitcoins was a way to buy and sell *other* things for Bitcoins. That, as Satoshi wrote to Martti, was the critical thing needed for enabling Bitcoin to catch on: "Not saying it can't work without something, but a really specific transaction need that it fills would increase the certainty of success."

The first, rather timid thrust in this direction was made by NewLibertyStandard in a post on the new Bitcoin forum:

What would you buy or sell in exchange for Bitcoins?
    Here's what I will buy if the price is right.
    Paper bowls, about 10 ounces (295 ml), no more than 50 count factory sealed.
    Plastic cups, about 16 ounces (473 ml), no more than 50 count, factory sealed.
    Paper towels, preferably regular size Bounty Thick and Absorbent, single roll, factory sealed.

Another user wondered what kind of wild celebration NewLibertyStandard was planning with all that disposable plate ware. "Bachelorhood?" NewLibertyStandard wrote back.

Soon thereafter, NewLibertyStandard began a Swap Variety Shop on his exchange website. Its selection was limited to a few sheets of postage stamps and SpongeBob SquarePants stickers.

Given this activity, it was not surprising that NewLibertyStandard soon shut down his exchange, while the network stagnated. Indeed, despite the recent innovations, at various points during late 2009 and early 2010 it appeared that the amount of computing power on the network was shrinking.

In the spring, Martti himself had less time to dedicate to the project after he dropped out of school and took a short-term, entry-level IT job with Siemens. Satoshi also went missing.

When Martti checked back in with Satoshi, in May 2010, he wrote, "How are you doing? Haven't seen you around in a while."

Satoshi's response was vague: "I've been busy with other things for the last month and a half—I'm glad you have been handling things in my absence."

In May a potential new user wrote to the Bitcoin mailing list, inquiring about how to accept Bitcoin for his web-hosting business. Sometime later he wrote again: "Wow, not one response in months. Amazing."

Another participant on the list, one of the first skeptics to criticize Bitcoin back in the fall of 2008, now wrote to explain: "Yes— Bitcoin kind of went dead."

He recalled the early debates on the cryptography mailing list with Satoshi about Bitcoin: "Long ago, I had an argument with the guy who designed it about scaling. I heard no more of it—of course with no one using it, scaling is not a problem. I do not know if the software is in usable condition, or has been tested for scalability."

But the apparent lack of activity in certain parts of the Bitcoin ecosystem obscured the fact that at a slow but steady rate it had been attracting a tiny but increasingly sophisticated core of users who were easy to miss if you didn't look carefully.

CHAPTER 4

●

April 2010

Laszlo Hanecz, a Hungarian-born twenty-eight-year-old software architect who lived in Florida, heard about Bitcoin from a programming friend he'd met on Internet relay chat, known as IRC. Assuming it was some scam, Laszlo poked around to figure out who was secretly making money. He soon realized there was an interesting and high-minded experiment going on and decided to explore further.

He began by buying some coins from NewLibertyStandard and then building software so that the Bitcoin code could run on a Macintosh. But like many good coders, Laszlo approached a new project with a hacker's mind-set, probing where he might break it, in order to test its robustness. The obvious vulnerability here was the system for creating, or mining, Bitcoins. If a user threw a lot of computing power onto the network, he or she could win a disproportionate amount of the new Bitcoins. Although Satoshi Nakamoto had designed the mining process so that the hash function contest would become harder if computers were winning the mining race more frequently than every ten minutes, those users with the most powerful computers still had a much better chance of winning a majority of the coins.*

Until now, no one had an incentive to throw lots of computing power into mining, given that Bitcoins were worth essentially nothing. But Laszlo decided to test this vulnerability. He understood that everyone on the network was trying to win the computational race with the central processing unit, or CPU, in his or her computer. But the CPU was also running most of the computer's other basic systems, so it was not particularly efficient at computing hash functions. The graphics processing unit, or GPU, on the other hand, was custom-designed to do the kind of repetitive problem solving necessary to process images and video—similar to what was needed to win the hash race function.

Laszlo quickly figured out how to route the mining process through his computer's GPU. Laszlo's CPU had been winning, at most, one block of 50 Bitcoins each day, of the approximately 140 blocks that were released daily. Once Laszlo got his GPU card hooked in he began winning one or two blocks an hour, and occasionally more. On May 17 he won twenty-eight blocks; these wins gave him fourteen hundred new coins that day.

Satoshi knew someone would eventually spot this opportunity as Bitcoin became more successful and was not surprised when Laszlo e-mailed him about his project. But in responding to Laszlo, Satoshi was clearly torn. If one person was taking all the coins, there would be less of an incentive for new people to join in.

"I don't mean to sound like a socialist," Satoshi wrote back. "I don't care if wealth is concentrated, but for now, we get more growth by giving that money to 100% of the people than giving it to 20%."

———————

* For more information on the mining process, see the Technical Appendix on page 357.

As a result, Satoshi asked Laszlo to go easy with the "high-powered hashing," the term coined to refer to the process of plugging an input into a hash function and seeing what it spit out.

But Satoshi also recognized that having more computing power on the network made the network stronger as long as the people with the power, like Laszlo, wanted to see Bitcoin succeed. Bitcoin's consensus model, which demanded that any new additions to the blockchain—and any changes to the Bitcoin software—had to be approved by a majority of the computers or nodes on the network, ensured that even if people tried to change the rules, or screw up the blockchain, they could not succeed without support from 50 percent of the other computers on the network. This model did leave the network vulnerable if one person or group captured more than 50 percent of the computing power, in what was referred to as a 51 percent attack. If Bitcoin supporters like Laszlo could add lots of computing power, that would make it harder for a bad guy to build up more than 51 percent of the power. And Laszlo did have the network's best interest in mind. It became clear on the forums that he was a good-natured guy and more interested in ideas than in personal wealth or success. Indeed, as he mined coins, he was eager to show how Bitcoin could be used in the real world. He posted in the forum asking if anyone would bake or buy him a pizza, delivered to his home in Jacksonville, Florida.

What I'm aiming for is getting food delivered in exchange
for Bitcoins where I don't have to order or prepare it
myself, kind of like ordering a "breakfast platter" at a
hotel or something, they just bring you something to eat
and you're happy!

Having stockpiled about 70,000 Bitcoins by this time, he offered 10,000 for a pizza. For the first few days no one accepted them. After all, what would the person on the other end do with the coins once Laszlo sent them over? But on May 22, 2010, a guy in California offered to call Lazlo's local Papa John's. A short while later a deliveryman knocked on the door of Laszlo's four-bedroom home in suburban Jacksonville bringing two pizzas, fully loaded with toppings.

Laszlo subsequently found several takers for the deal, which meant that for a few weeks he ate nothing but pizza. His two-year-old daughter was in heaven as he watched his stockpile of Bitcoins dwindle. But he had demonstrated that Bitcoins could be used in the real world. When he posted pictures from one of his feasts Martti Malmi cheered: "Congratulations laszlo, a great milestone reached."

LASZLO HAD PROVED that it was possible to pay for real things with Bitcoins, but the technology was still essentially just a vol-unteer project that relied on the goodwill of users. Perhaps the most notable project set up during these months was the Bitcoin faucet, a site that gave five free Bitcoins to anyone who registered. The project's creator was Gavin Andresen, a Massachusetts-based programmer who had spent $50 to get the 10,000 Bitcoins he was giving away, and who would become an almost mythic figure within Bitcoin. He first heard about the technology in May from a small item on the website of *InfoWorld*. After setting up the faucet, Gavin acknowledged that it sounded silly to give Bitcoins away, particularly because they were not hard to generate. But, Gavin wrote on the forums, "I want the Bitcoin project to succeed, and I think it is more likely to be a success if people can get a handful of coins to try it out. It can be frustrating to wait until your node gen-erates some coins (and that will get more frustrating in the future), and buying Bitcoins is still a little bit clunky."

Gavin, a trim forty-four-year-old with the anodyne looks of a suburban soccer dad, had time for the project because he, his two children, and his wife—a geology professor—had recently returned from his wife's sabbatical in Australia. Gavin had quit his job as a researcher at the University of Massachusetts before they had gone to Australia and he was now trying to figure out what to do next from his home office, just off the family mudroom.

When he first read about Bitcoin, he had immediately ferreted out Satoshi's original Bitcoin article, now known as the Bitcoin white paper, as well as the Bitcoin forum, all of which he read in a few hours. The concept appealed to him, in part, for the same political reasons that drew in Martti. After growing up in a liberal West Coast household, Gavin had moved toward libertarianism during his first programming job, swayed by a persistent coworker. These politics gave him a natural interest in a free-market currency like Bitcoin.

But politics didn't occupy the center of Gavin's life and, unlike many libertarians, he didn't particularly think the gold standard was a great idea. For Gavin, one of the primary attractions of this technology was the conceptual elegance of the decentralized network and the open source software, which was updated and maintained by all of its users instead of one author. Gavin's programming career thus far had given him an appreciation for decentralized systems that had nothing to do with any suspicion of the government or corporate America. For Gavin, the power of decentralized technology came from the more workaday benefits of software and networks that didn't rely on a single person or company to keep them running.

Decentralized systems like the Internet and Wikipedia could harness the expertise of all their users, unlike the AOL network or *Encyclopaedia Britannica*. Decision making could take longer, but the ultimate decisions would incorporate more information. The participants in decentralized networks also had an incentive to help keep the system up and running. If the original author was away on vacation or asleep when a crisis hit, other users could chip in. As it was frequently put, systems were stronger when there was no single point of failure. These arguments were, to some degree, technological analogues of the political arguments that libertarians made for taking power away from central governments: political power worked better when it was in the hands of lots of people rather than a single political authority. But the advocates for open source software tended to put things in less ideological terms.

Decentralized technology was a rather natural fit for Gavin, who had little in the way of an ego. Despite going to Princeton, he had been happy serving as something of a journeyman programmer, working on 3-D graphics at one point, and Internet telephony software at another. For Gavin, the jobs had always been about what he found interesting, not what promised the most money or success.

To start participating in the Bitcoin project, Gavin quickly began e-mailing with Satoshi to suggest his own improvements to the code and, in short order, became the first person other than Satoshi or Martti to officially make a change to the Bitcoin code.

More valuable than Gavin's programming chops were his goodwill and integrity, both of which Bitcoin desperately needed at this point to win the trust of new users, given that Satoshi remained a shadowy figure. Satoshi had, of course, designed his software to be open source so that users wouldn't have to trust him. But people were not showing much willingness to entrust real money to a network that was run by a bunch of anonymous malcontents.

Gavin attached a real and trustworthy face to the technology. He was one of the first people on the forum to use his real identity, taking the screen name gavinandresen, and he included, on the forum, a small picture of himself in a hiking backpack, giving a slightly dorky but entirely disarming smile. He served on the forums as a sort of good-natured high school teacher, answering, in plain terms, questions that came up. He would also mediate in the political fights that occasionally broke out between those early users with strident political beliefs. Gavin was used to this sort of thing. In Amherst, Massachusetts, he served on the 240-member Town Committee, a grassroots deliberative body that he had been elected to a number of times. Amherst, a college town, was famously liberal and so Gavin had plenty of disagreements over matters of principle. But he had learned to avoid fights and find compromises—something that was about to prove critical to the fledgling Bitcoin community.

HEADPHONES ON AND an oversize can of MadCroc energy drink by his side, Martti sat at his dorm room desk, giddy. Slashdot, a go-to news site for computer geeks the world over, was going to post an article about Martti's pet project. Bitcoin, largely ignored over the last year, was on the verge of receiving global attention.

The campaign to get Bitcoin real press coverage had begun a few weeks earlier, not long after Martti finished his three-month internship at Siemens. A new version of Bitcoin, version 0.3, was being prepared for release by Satoshi, and the regulars on the forum saw a perfect opportunity to get the word out. Martti agreed with a handful of other users that Slashdot would be the best place to do this.

"Slashdot with its millions of tech-savvy readers would be awesome, perhaps the best imaginable!" Martti wrote on the forum. "I just hope the server can stand getting 'slashdotted.'"

A small crew went back and forth about the right language to submit to the Slashdot editors. Satoshi got his hackles up when someone suggested Bitcoin be sold as "outside the reach of any government."

"I am definitely not making any such taunt or assertion," Satoshi wrote.

He quickly apologized for being a wet blanket: "Writing a description for this thing for general audiences is bloody hard. There's nothing to relate it to."

After Martti suggested his own changes, the final version made the more modest assertion that "the community is hopeful the currency will remain outside the reach of any government."

When the item went online, shortly after midnight in Helsinki, it wasn't anything more than the single paragraph the Bitcoin team had submitted.

"How's this for a disruptive technology?" it began. "*Bitcoin* is a peer-to-peer, network-based digital currency with no central bank, and no transaction fees."

Despite the modesty of the item, the Internet chat channel that Martti had established for the Bitcoin community quickly lit up. NewLibertyStandard wrote: "FRONT PAGE!!!"

Regulars like Laszlo made a point of being on the Bitcoin chat channel, to answer questions and serve as a tour guide of sorts for any newbies who checked in after reading the story. In his dorm room, Martti posted a message on Facebook: "If I was a smoker, I would have smoked two packs already."

Martti watched as the counters, which tracked the number of users on the forum and the chat channel, ticked steadily upward. Messages crowded his forum in-box; and the Bitcoin website, running on servers that could not handle more than one hundred viewers at a time, began to slow. Within an hour, the limit was reached and the whole site went down. Martti scrambled to scale up the site's capacity with the company that rented him space. But this, and the derogatory comments that showed up under the Slashdot item, did not dampen his enthusiasm. This was what he'd been waiting for for months.

CHAPTER 5

●

July 12, 2010

When he awoke late, the morning after the Slashdot posting, Martti Malmi saw that the attention was not a hit-and-run phenomenon. People weren't just taking a look at the site and moving on. They were also downloading and running the Bitcoin software. The number of downloads would jump from around three thousand in June to over twenty thousand in July. The day after the Slashdot piece appeared, Gavin Andresen's Bitcoin faucet gave away 5,000 Bitcoins and was running empty. As he begged for donations, he marveled at the strength of the network:

> Over the last two days of Bitcoin being "slashdotted" I
> haven't heard of ANY problems with Bitcoin transactions
> getting lost, or of the network crashing due to the load, or
> any problem at all with the core functionality.

But while the Bitcoin software itself was working well, new users quickly ran up against the limitations of the Bitcoin ecosystem. Those who immediately wanted to acquire more Bitcoins than were available from Gavin's faucet were left with only a few meager options, one of them a creaky, unreliable service that Martti had set up a few months earlier.

Jed McCaleb was one of the people who encountered this weakness. A native of Arkansas, Jed had been raised by his single mother, who made a living as a journalist. From a young age, Jed had been something of a math and science prodigy, and this allowed him to make it to Berkeley for college. Jed, though, had trouble sticking with things, and he soon dropped out of Berkeley and moved to New York. There he and a partner set up what became one of the main successors to Napster. His software, eDonkey, made it possible for individuals to trade large files like movies and it proved so successful that the Recording Industry Association of America sued Jed and his business partner. They eventually paid $30 million to settle the case and shut eDonkey down, but they also earned a few million along the way.

Despite being a soft-spoken introvert, Jed had a cool way about him that helped him make friends and girlfriends. When one of his romantic flings ended up pregnant, he and the woman, MiSoon, decided somewhat spontaneously to keep the baby and make a go of it. They used some of Jed's earnings to buy an estate with a pool an hour or so north of New York City, just as they were expecting a second baby. In the sprawling, mostly empty house, Jed threw himself into an online game he had created called *The Far Wilds*, which had attracted only a few aficionados. He spent endless hours in a first-floor bedroom, which he had turned into a den. Books about neuroscience and artificial intelligence piled up around him—as did old food, attracting bugs that MiSoon initially tried to get rid of, but later came to accept as one of the side effects of Jed's brilliant mind.

When Jed came across the Slashdot post about Bitcoin he was immediately intrigued. It seemed to fulfill many of the ideals behind Napster and eDonkey—taking power from authorities and giving it to individuals. But when Jed tried to buy some actual Bitcoins, he ran into the limitations of the few existing sites that sold them.

MiSoon was nursing their newborn son when she wandered into Jed's study one night and encountered his frustration.

"There's this really cool thing called Bitcoin—it's like this nerd, libertarian thing," Jed told MiSoon, in his hushed, intense voice. "But it's so lame. I can't buy any at night."

Jed said he wanted to build a site himself where he could buy coins at any hour. When MiSoon arose the next morning, it was done. With some experience in amateur foreign-currency trading, Jed knew the basics of what an exchange required. But he had never actually set up a website before, having previously worked more on the sophisticated back-end software. His new Bitcoin exchange was something of a fun experiment.

He and MiSoon discussed possible names for the site. He mentioned an old domain name that he owned and was not using—mtgox.com. Jed had bought the site in 2007, for use as an online exchange to buy and sell the cards used in the role-playing game *Magic: The Gathering*—hence the acronym for *Magic: The Gathering Online Exchange*. It had operated for just a few months before Jed shut it down and the site had been vacant since.

"Yeah, you should use that," MiSoon replied. "That's kind of weird and easy to remember. Why not if you already have it registered?"

Seven days after the Slashdot post, Jed casually advertised his new site on the Bitcoin forum:

Hi Everyone,
I just put up a new Bitcoin exchange.

Please let me know what you think.

Mt. Gox was a significant departure from the exchanges that already existed, primarily because Jed offered to take money from customers into his PayPal account and thereby risk violating the PayPal prohibition on buying and selling currencies. This meant that Jed could receive funds from almost anywhere in the world. What's more, customers didn't have to send Jed money each time they wanted to do a trade. Instead, they could hold money—both dollars and Bitcoins—in Jed's account and then trade in either direction at any time as long as they had sufficient funds, much as in a traditional brokerage account.

These advances made it significantly more convenient to buy and sell Bitcoins, but also brought new dangers that threatened to betray some of the currency's basic principles. Satoshi had designed Bitcoin to eliminate the need for trusted central authorities. It was supposed to be a new money that people could hold on their own, without a bank, secured with a private key that only the user knew. Mt. Gox customers would be moving back to the old model in which a single institution—Jed's company—held everyone's money. If Jed offered good security measures, this might prove safer than holding coins on a home computer. But Jed was not a security expert, and if he did somehow lose the private keys to the exchange's digital wallets, his customers had little recourse. Unlike the banks that Bitcoiners had bashed, Mt. Gox had no deposit insurance and no regulators overseeing the safety and soundness of Jed's operation. The choice was between security and principles on one hand and convenience on the other.

When a forum member asked why they should choose Mt. Gox over the alternatives, Jed responded in his characteristically modest but confident way.

"It is always online, automated, the site is faster and on dedicated hosting and I think the interface is nicer."

Even Jed, though, was surprised at how quickly people trusted his setup and sent money to his PayPal account. During his first day in business, July 18, twenty Bitcoins were traded at five cents each on Mt. Gox—an inauspicious opening. But within the first week he had his first hundred-dollar day of trading, and by the end of the month Mt. Gox had overtaken Martti's service and the other existing exchange in trading volume to become the largest Bitcoin business around.

These weeks marked a subtle but dramatic transition for Bitcoin. Until this point, there had been occasional transactions, but mostly between aficionados making them out of a desire to help the network. After the Slashdot story, the difficulty of mining new Bitcoins ramped up quickly with the surge in the number of people racing to win coins. Satoshi had determined that as more computers joined the network, the mining of new Bitcoins would become more difficult, ensuring that it would always be roughly ten minutes between releases of new coins. The week after the Slashdot story, the difficulty of mining new Bitcoins jumped 300 percent. Gavin Andresen, who had initially started mining Bitcoins to help the network, now found it all but impossible to win new coins with his four-year-old Mac laptop.

Suddenly, if a person wanted Bitcoins, he or she had to buy them. And people were showing a willingness to do just that and part with real money for these unproved slots on a digital spreadsheet. The growing popularity of Bitcoin was hard to miss. One new forum member wrote:

What I like about Bitcoin is that it is a community with a solution that we are actually trying. I don't know many people in real life that are even close to as radical in their thinking as I (and many others on these forums) am. Surprisingly, however, I am able to talk with my real life friends about Bitcoin much longer than my normal rants about "what should be," because Bitcoin actually exists.

IN LATE JULY Martti launched the first foreign-language forum, in Russian, and within a few weeks it had hundreds of postings. The English forum grew much faster. In one month, the forum had gained more new members—370—than it had since coming online in November 2009. Craving more conversation, the expanding herd of dedicated Bitcoin followers found their way to the chat channel Martti had set up. Now, the Bitcoin channel on Internet relay chat, or IRC, became a sort of twenty-four-hour global coffeehouse where the new users could gather and marvel at this experiment they were all taking part in.

Around midnight on September 26, one new Bitcoiner wrote: "gosh I can't sleep ! I keep thinking about this great stuff. To me Bitcoin is the 'cyberspace gold.' I'm just amazed."

The next afternoon another new user spoke of spending ten hours reading everything he could find about the network.

"I did the same thing when I first heard about Bitcoin," Gavin wrote back.

The appeal of Bitcoin varied from person to person, but most were in love with the basic idea of a digital cash that each user could control and move around the world with nothing more than a private key. The users, at this point, were mostly young men whose lives were untethered to anything other than their laptops, in constant communication with people on the other side of the world. For them, moving money around the globe with a paper check or an old-fashioned wire transfer seemed absurdly backward.

Satoshi chimed in on the forums to note that the Bitcoin software was designed to do more than just move coins. The software also had the capability to attach specific instructions to each coin so that the coins could behave in a particular way, according to the users' wishes. A coin on the blockchain could, for example, be programmed to move from one address to another only if it was signed off on by three or four different private keys, enabling its use in the types of legal transactions that currently required cumbersome and expensive middlemen.

"The design supports a tremendous variety of possible transaction types that I designed years ago," Satoshi wrote. "Escrow transactions, bonded contracts, third party arbitration, multi-party signature, etc. If Bitcoin catches on in a big way, these are things we'll want to explore in the future, but they all had to be designed at the beginning to make sure they would be possible later."

Satoshi had advertised Bitcoin as a trustless system that didn't require its users to rely on any central authority. But like all forms of money, Bitcoin did rely on its users' trusting the ideas and integrity of the system supporting it—in this case, code and math—and the small elite of cosmopolitan coders was more than willing to do that. These new converts, in turn, were providing not just enthusiasm, but also fresh sets of eyes to examine the code with a level of programming experience that had been scarce up to this point.

In late July Gavin and Satoshi got an e-mail from one such user, a programmer from Germany going by the screen name ArtForz, who had found a previously undiscovered weakness in the code that governed transactions on the network. The flaw made it possible to spend Bitcoins in someone else's wallet.

Gavin and Satoshi immediately realized this was not just a bug but a fatal flaw that could doom the entire project. If someone else could spend your coins the whole system was all but useless.

Satoshi quickly put together a fix—the flaw was not actually difficult to correct. But in the meantime, Gavin and Satoshi agreed to keep the flaw secret until they got everyone on the network using new, repaired code, for fear that someone would take advantage of it.

"For now, don't call it the '1 RETURN' bug to anyone who doesn't already know about it," Satoshi wrote to Gavin.

Because the patched software "has a dozen changes in it," Satoshi wrote, "it won't necessarily be obvious what the worst vulnerability was. That may give people a head start to upgrading if any attackers are looking for the vulnerability in the changes."

That ArtForz had not taken advantage of the bug himself was a minor miracle. But it was also what the incentives in the Bitcoin system were designed to encourage. ArtForz had been mining coins himself—using the GPU technology that Laszlo had first pioneered—and he knew that if confidence in the system was undercut his coins would be worthless. The market incentives were working as they were supposed to work. This turn of events also confirmed Gavin's confidence in the power of decentralized systems. ArtForz was a part of the network, and as such, he didn't just passively use the network. He and Gavin, and all the others, were helping to build this thing.

A FEW MONTHS earlier the big concern plaguing the Bitcoin forum was how to attract new users, but now the problem was how to deal with the influx of new users, their potentially malicious behavior, and their competing interests.

These problems became particularly pronounced after Bitcoin's next big jump into the spotlight. In November, WikiLeaks, the organization founded by a regular participant in the old Cypherpunk movement, Julian Assange, released a vast trove of confidential

American diplomatic documents that revealed previously secret operations around the world. The large credit card companies and PayPal came under immediate political pressure to cut off donations to WikiLeaks, which they did in early December, in what became known as the WikiLeaks blockade.

This move pointed to the potentially troubling nexus between the financial industry and the government. If politicians didn't like the ideas of a particular group, government officials could ask banks and credit card networks to deny the unpopular group access to the financial system, often without requiring any judicial approval. The financial industry seemed to provide politicians with an extralegal way to crack down on dissent.

The WikiLeaks blockade went to the core of some of the concerns that had motivated the original Cypherpunks. Bitcoin, in turn, seemed to have the potential to counteract the problem. Each person on the network controlled his or her coins with his or her private key. There was no central organization that could freeze a person's Bitcoin address or stop coins from being sent from a particular address.

A few days after the WikiLeaks blockade began, *PCWorld* wrote a widely circulated story that noted the obvious utility of Bitcoin in the situation: "Nobody can stop the Bitcoin system or censor it, short of turning off the entire Internet. If WikiLeaks had requested Bitcoins then they would have received their donations without a second thought."

It wasn't clear if Bitcoin could actually be used in this particular instance, but whatever the practical possibilities, the blockade was helping elevate the debate around Bitcoin beyond the rather narrow issues of privacy and government money-printing that had been dominant in the early days. Here was a broader philosophical issue that could attract a wider audience, and the forums were full of new members who had been drawn in by the attention. One

new user, a young man in England named Amir Taaki, proposed
making Bitcoin donations to WikiLeaks. Amir argued this could
raise Bitcoin's profile at the same time that it could help WikiLeaks
raise money.

This kicked off a vigorous debate on the forum. A number of
programmers worried that the Bitcoin network was not ready for
all the traffic—and government scrutiny—that might come if it
started to be used for controversial donations.

"It is extraordinarily unwise to make Bitcoin such a highly visible target, at such an early stage in this project. There could be
a lot of 'collateral damage' in the Bitcoin community while you
make your principled stand," one programmer wrote.

Satoshi eventually ended the debate. When someone on the
forum wrote, "Bring it on," Satoshi responded forcefully:

No, don't "bring it on."
    The project needs to grow gradually so the software
can be strengthened along the way.
    I make this appeal to WikiLeaks not to try to use Bit-
coin. Bitcoin is a small beta community in its infancy. You
would not stand to get more than pocket change, and the
heat you would bring would likely destroy us at this stage.

This was enough to convince Amir.

"I've done a U-turn on my earlier view and agree. Let's protect
and care for Bitcoin until she leaves her nursery onto the economic
killing fields."

This was one of an ever-diminishing number of communications from Satoshi during the fall of 2010. Messages from both
Satoshi and Martti had been increasingly rare. In Martti's case,
after a year of working on Bitcoin free, he needed a regular source
of income. In September, two months after the Slashdot story, he

took a full-time job with a firm that analyzed social-media data. On top of having a full schedule, Martti also saw that he was no longer needed. Gavin and a few others were taking over many of the day-to-day tasks that Martti had previously handled. And the chat channels were crawling with people ready to help out.

Satoshi's gradual fading was less explicit. He still posted occasionally to the forums when there were specific questions, but he never appeared on the chat channel and increasingly shifted to infrequent private communications with Gavin and just a few other developers. In December, Satoshi asked Gavin if he would mind having his e-mail address posted on the Bitcoin website, as a point of contact. After his own name went up, Gavin noticed that Satoshi's e-mail came down.

When the last public forum post came from Satoshi, on December 12, 2010, there was nothing marking it as such. Announcing the latest version of the software, version 0.3.19, the post was markedly different in tone from those early messages, selling the world-beating potential of Bitcoin. The main sentiment now was a warning that Bitcoin was still extremely susceptible to denial-of-service attacks, which overwhelm a system with message traffic.

"There are still more ways to attack than I can count," Satoshi wrote in the brief note.

This came just days after Hal Finney checked back in for the first time since early 2009. His disease, ALS, had progressed quickly and he was now largely confined to the family living room, in a special setup his family had concocted so that he could continue working on a computer.

Hal made an unassuming return to the community with some relatively dry comments about patterns in the price of Bitcoin and the possibility of using Bitcoin's blockchain as a new kind of database. He was as enthusiastic as ever about the network.

"I'd like to hear some specific criticisms of the code. To me it looks like an impressive job, although I'd wish for more comments," he wrote on the forum. "This is some powerful machinery."

This provoked Satoshi's second-to-last post: "That means a lot coming from you, Hal. Thanks."

This exchange set off a discussion among people who had never heard Hal's name before.

"Who is Hal on the forum?" one user wrote. "Satoshi seemed to know of him."

The question quickly gave way to the bigger mystery: Who is Satoshi?

"Is he a real person? ;-)" a forum user asked.

"Hmm, there are almost no results for Satoshi unrelated to Bitcoin," another user wrote after some quick research.

This set off the first stages of a hunt for Satoshi that would continue for years. People on the chat channel began debating the available details about Satoshi and their significance. It was noted that Satoshi occasionally used British spellings and words like "bloody." There was also a fragment from a British news story written into the first block of Bitcoins created by Satoshi's computer.

A Bitcoin user in Japan noted that Satoshi was a common name in Japan, but he argued that Satoshi was unlikely to be Japanese given that Satoshi had never used Japanese words and had always written his name with the family name last, contrary to Japanese tradition.

"Maybe this is a gambit to trick us to think he's not Japanese," another user wrote.

"I like the pseudonym theory the best. It's so much cooler for someone to have a secret identity than just a boring name," someone wrote.

"Jesus, this is a great story. I'm amazed the NY Times hasn't picked up on it yet," another poster chimed in.

In the early days, Martti had never asked Satoshi any personal questions but had assumed that Satoshi Nakamoto was probably not a real name. Martti's access to the Bitcoin websites allowed him to see that Satoshi was joining the sites through a Tor network that obscured his geographic location and IP address.

Gavin had asked Satoshi some personal details in his first e-mail, but Satoshi ignored the questions and Gavin never pressed for more.

One regular forum user asked Satoshi: "Suppose, god forbid, you were no longer able to program or were unavailable due to unknown circumstances. Do you have a procedure in mind to continue Bitcoin in your absence?"

Satoshi didn't answer, but others on the forum noted that because Bitcoin's software was open source, available to all the users, Satoshi's involvement shouldn't matter: "As long as the source code remains open, that is sufficient. If there is a need, and enough interest, the community will provide. Trust in the community :)" one developer wrote.

Satoshi was, in many ways, just as powerless, or powerful, as every other user on the network. All the coins were on the communal blockchain, but only the person with the private key corresponding to each address on the blockchain could use the coins in that address. Satoshi could try to change the software in some way that would give him more control, but doing so wouldn't gain traction unless a majority of the network adopted the changes.

Still, Gavin, who was now perhaps the most central figure in Bitcoin, knew that the platonic ideal of open source software was somewhat more complicated underneath the surface. While anyone could propose changes to the protocol, he and Satoshi were still essentially the only people who could sign off on changes—and this gave them an unusual amount of power in the system. What's more, while Satoshi had written a program designed to eliminate the need for trust, users of the technology still had to have faith that it would work as intended. On the forum, Gavin wrote: "Trust is Bitcoin's biggest barrier to success. I don't think there is anything we can do to speed up the process of getting people to trust that Bitcoin is solid; it takes time to build trust."

At this point, though, the primary cause for distrust was not the lack of information about Satoshi. Satoshi's anonymity, if anything, seemed to increase the level of faith in the system. The anonymity suggested that Bitcoin was not created by a person seeking personal fame or success. What's more, Satoshi's absence allowed people to project their own vision onto Bitcoin.

Those who could cause problems, though, were the very people who were making Bitcoin grow. The network was expanding, but the people among its growing ranks would also pose the greatest threat to Bitcoin and the trust it needed.

CHAPTER 6

●

September 2010

The Sony Vaio laptop that was the nerve center of the biggest business in the Bitcoin world in the fall of 2010—Mt. Gox—sat on a square wooden table, under a roof made out of dried palm leaves. An oblong swimming pool was just feet away.

The founder of Mt. Gox, Jed McCaleb, had moved to Nosara, a Costa Rican beach town, less than two months after starting the exchange. Lonely in their isolated New York estate, he and MiSoon didn't want to spend another winter cooped up with their two small children. In Nosara they found a house near the beach, with a Montessori school for the children, an opportunity for Jed to finally perfect his surfing, and a hut in the backyard where he could work.

But the booming new business was not cooperating with their plans for a quiet tropical life. Just ten days in, he had seen his first day with 1,000 Bitcoins traded and about ten days after that he saw his first day with over 10,000 Bitcoins traded, meaning that over $1,000 changed hands that day. Jed was making 0.5 percent from each side of every trade, a nice reward for something that required little work. But the flow of money in and out, particularly from PayPal, was causing headaches.

Jed suffered from an issue common in any business that takes credit cards or PayPal. All the traditional payment networks allow customers to dispute charges and can take money back from merchants, like Jed, even after transactions go through. This was one of the issues that Cypherpunks had wanted to address in creating digital cash—owing to the anger about how much power the system of so-called chargebacks gave to the credit card companies of the world. Bitcoin itself did not allow charges to be reversed, but if Jed sold Bitcoins via PayPal to someone who then disputed the PayPal payment, Jed could lose the PayPal money and not be able to get the Bitcoins back. Within a month, Jed acknowledged he was defenseless against this.

"I'm just eating the charge which sucks so please, please don't do this," he pleaded on the forum.

After this post, the problem got worse, not better. Jed tried to resolve disputes before they escalated, even if it meant losing money, so he didn't have his PayPal account shut down altogether. But one morning he opened up his laptop and found that PayPal had done just that, leaving him without an easy way to get money from customers. Meanwhile, people who had money stuck in Jed's frozen PayPal account complained about the difficulty of getting it back.

"I do this in my spare time for free so don't get all uppity," Jed wrote to his critics.

This was clearly not what Jed signed up for when he opened Mt. Gox. He had never intended for it to become a full-time job. He was motivated by working on interesting challenges, and Mt. Gox was instead becoming a series of boring and stressful problems. Like many people interested in big challenges and bold solutions,

Jed got bored by the details of seeing those solutions to their end—
something that would come back to haunt the community later.

On the hunt for someone who could help relieve him of the
burden of work on Mt. Gox, Jed began chatting online with a
user named MagicalTux, whom Jed soon came to know as Mark
Karpeles. Mark was almost always online because it was one of
the only places where he felt comfortable in the world. A chubby
twenty-four-year-old, Mark had been raised in France alternately
by his mother and grandmother, who didn't get along and contin-
ually moved him between schools. At age ten, Mark was sent to a
Catholic boarding school in the Champagne region of France—a
school he looked back on with fear and anxiety. Even as a youngster, Mark had tremendous difficulty with human interaction,
while the logic of the computer had spoken to him naturally. He
would ace his math classes—and could assemble and disassemble
his calculators—but he struggled with literature and the human-
ities, and eventually dropped out of school, not long before he was
arrested for some of his hacking activities. Since then, he'd had
a peripatetic lifestyle, looking for a place where he could feel at
home. He first tried Israel, thinking it might help him get closer to
his Catholicism, but he soon felt as lonely as ever, and the servers he
was running kept getting disrupted by rocket fire from Gaza. Back
in France, he got a job as a programmer but soon fell out with his
boss. During this period, he would make rather melancholy posts
to a generally unread blog in which he discussed his situation.

"To tell the truth, I always felt a sort of emptiness in my existence, somewhat as if I wasn't really in the right place, or as if I
was missing something I needed in order to really live, and not just
survive," he wrote in 2006.

Mark finally got a chance to visit Japan, which he had been
drawn to since reading a series of Manga comics his mother had

given him. When he arrived the first time and checked into his capsule hotel, the part of him that had always been afraid in France was put to rest by the stoicism and politeness of Japanese culture. It didn't hurt that the girls in Japan seemed to actually respect the fact that he was a programmer.

By the time he met Jed online, Mark had lived in Tokyo for more than a year and set up his own web-hosting company that rented out server space. He learned about Bitcoin from a French customer in Peru who wanted an easier way to pay the bills Mark sent him. As Mark dived into Bitcoin in late 2010, he discovered that it had already attracted an unusually cohesive and friendly online community, the sort of social setting in which he could feel comfortable. He would engage in endless chats at all hours about everything from obscure Japanese payments systems to the identity of Satoshi, who Mark was confident was not Japanese.

"I'm a coder and already worked with tons of japanese people here, and the way the code is made is also completely different from anything I ever saw in japan (but not so different from more western stuff)," Mark wrote one night on the chat channel.

Online, Mark had a brash cockiness that he never showed in real life—so brash, in fact, that it was occasionally off-putting. But he lived alone with his cat, Tibanne, and was always available and willing to help out. He volunteered to help Martti Malmi host the Bitcoin website on his servers. And when Martti offered to connect Jed with his European bank, so Mt. Gox could begin accepting euros, Mark helped Jed set up the back end. The work gave Jed confidence in Mark's abilities.

As the price of Bitcoin rose to nearly 30 cents per coin by the end of December 2010—thanks, in no small part, to the attention from WikiLeaks—Jed called a lawyer in New York to ask about the regulatory implications of running a business like Mt. Gox. The lawyer said it was unclear how the government would view

Bitcoin. In the forums, there were lengthy debates about whether Bitcoin would be considered money, which would be subject to bank regulators, or some sort of commodity, which would come under different government oversight. Whatever the outcome, the lawyer told Jed that he would probably have to eventually register as a money-transmission business, which would involve extensive applications and lots of legal bills.

Jed turned to Mark for advice, seeking his thoughts on a four-page document Jed had put together to send to potential investors. The document underscored how far Mt. Gox had risen in its short life. The business was worth $2 million by Jed's estimate: "Mt. Gox is generating revenue with very low running costs and huge potential upside," the document said. Jed told Mark he was thinking of raising about $200,000, mostly to hire a lawyer to help deal with the regulatory situation.

But as the headaches continued to pile up, Jed got more antsy. In January, a Mt. Gox user named Baron managed to hack into Mt. Gox accounts and steal around $45,000 worth of Bitcoins and another type of digital currency that Jed had been using to transfer money around. When Baron deposited $45,000 back into Mt. Gox to buy more Bitcoins, Jed froze Baron's money. The incident reinforced Jed's belief that Mt. Gox was a prime target for hackers and that he had neither the time nor the security expertise to protect it adequately.

Jed wrote to Mark: "Please keep all this confidential. I don't want to start a panic, and I'm not sure I'll do it yet, but I'm thinking I might try to sell Mt. Gox."

When Mark picked up the conversation on the Internet relay chat (IRC), Jed asked if Mark would be interested in purchasing the site and made him an offer that was hard to refuse. Mark would not have to pay anything up front. All he would have to give up was 50 percent of the company's revenues for the first six months. Jed would continue to hold 12 percent of the company, but Mark could have the rest. Jed's fraction of the company was designed to be small enough to protect him from legal liability if Mt. Gox ran into problems in the future.

Jed and Mark were outwardly very different people. Mark was a large, awkward Frenchman, while Jed was a slight, suave American. But both of them were loners who tended to skeptically watch the world from afar and live mostly in their own heads. Each was the only child of a single mother who had given him self-confidence while also making him skeptical about traditional sources of authority—a mixture of traits that made for a good match with Bitcoin at this point.

As the deal between the two men progressed, the strange legal limbo in which Bitcoin existed colored every step. Neither Mark nor Jed used a lawyer. Instead they drew up contracts themselves and sent them back and forth. After they had both signed these contracts, Mark wrote up a less-than-official-looking certificate that said that Jed officially owned forty shares of Mt. Gox, though it did not say how many total shares existed.

Jed didn't labor over the deal because, even with all the growth Mt. Gox had experienced, the business still had fewer than three thousand customers, and was on track to bring in only around $100,000 in revenue for the year.

Mark took ownership of Mt. Gox using the corporation that also held his web-hosting business, Tibanne Ltd.—named after his orange-and-white tabby cat.

By the time Mark and Jed finished their deal, the price of Bitcoin had shot above $1, attracting a new wave of media attention. It also attracted another big hacking attack. At this point, of the 21 million Bitcoins that would ever be released, one-fourth were now out in the world, worth around $5 million at the $1 exchange rate. What's more, the number of daily transactions was creeping steadily upward.

The cause of this surge was due, in no small part, to the rise of another business that was to pose an even graver test to the foundation of trust that Bitcoin was trying to build.

THE POSSIBILITIES FOR using Bitcoin in the real world had not progressed much since NewLibertyStandard's offer of SpongeBob SquarePants stickers. Mark Karpeles was still taking Bitcoin for his web-hosting services and a farmer in Massachusetts was selling alpaca socks. But the range of products available for Bitcoin expanded in a dramatic way a few days before the price of Bitcoin shot from around 50 cents to above $1 for the first time, when an unassuming post on the Bitcoin forum heralded the next wave of Bitcoin commerce.

"Has anyone seen Silk Road yet? It's kind of like an anonymous amazon.com. I don't think they have heroin on there, but they are selling other stuff."

The posting was made by someone who went by the screenname altoid. In real life, he was Ross Ulbricht, a 6-foot-2 surfer-cum-scientist who had been planning Silk Road for months when he put his innocent-sounding post on the forum.

For Ross, a fun-loving, well-educated twenty-six-year-old, the creation of Silk Road had begun in earnest in July 2010 when he had sold a cheap house in Pennsylvania that he'd acquired while he was a graduate student there. With the $30,000 from the sale, Ross rented a cabin about an hour from his home in Austin, Texas. He also purchased petri dishes, humidifiers, and thermometers, along with peat, verm, gypsum, and a copy of *The Construction and Operation of Clandestine Drug Laboratories*, by Jack B. Nimble.

The psychedelic mushroom lab he set up in the cabin was not created with the intent of enabling Ross to become a petty drug dealer. He had much grander visions of his life than that. From the time he sold the house in Pennsylvania, he knew he wanted to set up a new kind of online market, where people could buy all the things that aren't available on ordinary online markets.

This unusual and dangerous business concept was the product of the idiosyncratic mixture of influences that had shaped Ross's mind. His parents had been hippies of sorts, taking him on vacations to Costa Rica, where his father taught him to surf. His curiosity about and penchant for the outdoors had later helped turn him into a seeker, looking for ways to free his mind and achieve oneness through Eastern philosophy and designer drugs. Ross came from Texas, and his search for freedom led him to some of the thinkers on the border between libertarian thought and anarchism—the same philosophers who had influenced many of the Cypherpunks—and he came to believe that the ultimate hurdle to personal freedom was government. At Penn State, he had the unique distinction of being a member of both the campus libertarians and the West African drumming ensemble. He would describe his ideological awakening in spiritual terms.

"Everywhere I looked I saw the State, and the horrible withering effects it had on the human spirit," Ross would say. "It was horribly depressing. Like waking from a restless dream to find yourself in a cage with no way out."

In Austin, Ross did not tell anyone about the new marketplace he was working on, but he did give some indication of what he was after on his LinkedIn page, where he wrote, in broad terms, that he was "creating an economic simulation to give people a first-hand experience of what it would be like to live in a world without the systemic use of force."

Initially, he called the project Underground Brokers, but soon enough he settled on a more enticing name: Silk Road. The mushrooms growing in the cabin were going to be just the first product, so something would be available for purchase when the site opened—and he soon had big black trash bags full of them.

In building Silk Road, the drugs were the easy part. The harder part was finding a way to sell the drugs online, outside the watchful gaze of the authorities. The first necessary tool he'd discovered was software, known as Tor, which allowed people to obscure their location and identity when surfing the Web. It also allowed for websites to be set up behind a similar curtain of anonymity. While Tor had been created by United States Naval Intelligence, to give dissidents and spies a way to communicate, it was based on ideas that had been developed by David Chaum and other cryptographers. Most Tor websites could be visited only by people using a Tor web browser. The web address that Ross posted on the Bitcoin forum for Silk Road—http://tydgccykixpbu6uz.onion—gave it away as a Tor site.

The second important tool that Ross had discovered was Bitcoin. With Tor alone, a customer wanting to buy Ross's mushrooms could have visited Silk Road without being tracked. But assuming the customer didn't want to pay by sending cash through the mail, all the other alternatives for making digital payments were easily tracked—as the Cypherpunks well knew. Ross saw that Bitcoin solved this problem. If a buyer paid for drugs with Bitcoin, the Bitcoin blockchain ledger would record coins moving, but the Bitcoin addresses on either end—a series of letters and numbers—would not include the names of the people involved in the transaction. Now the only identifying information about the buyer was the postal address where he or she asked to receive the drugs. And this was easy to game by providing anonymous post office boxes.

Within the Bitcoin world, there had been a common assumption that people looking to buy illegal or unsavory goods were likely to be among the first to have an incentive to use Bitcoin. In one early conversation about where Bitcoin might catch on,

Satoshi had argued for online porn, where users "either don't want the spouse to see it on the bill or don't trust giving their number to 'porn guys.'"

Ross had made his first post about Silk Road in the middle of a long-lasting thread on the Bitcoin forum, entitled "A Heroin Store," which had been discussing the possibility of such a marketplace. Martti had chimed in a few months earlier, helpfully trying to think of ways to make it work. For him, the sticking point was how to get both sides of the transaction to trust each other enough to part with their Bitcoins and drugs.

The fact that Ross had figured out how to put all the pieces together was a minor miracle. Ross had studied physics in college and materials science in graduate school at Penn State. But he was only an amateur programmer and he had to learn the nuances of Tor and Bitcoin software as he went along, stumbling at many points. His ability to pull it off was a testament to his work ethic and business acumen. In response to Martti's concern, he created an escrow service—essentially himself—to hold the Bitcoins of a customer until the drugs arrived in good condition, so the customer had some recourse if the pills or powder didn't show up as expected. On the programming front, Ross managed to sweet-talk an old college friend, who was a more experienced programmer, into giving him lots of technical advice.

In addition to all this, though, Ross's ability to get Silk Road up and running was a product of his sheer desperation at a difficult moment in his life. Two years earlier, Ross had abandoned graduate school—despite having already published several scientific papers—because he wanted to do bigger things with his life. The first things he tried all fell flat, including a used book store he was running at the time he put Silk Road online. This had been one of the first prolonged periods of struggle in a life that had otherwise been quite charmed. Ross had movie star looks that won him comparisons to the actor Robert Pattinson, and he had always had an easy time making friends, attracting women, having fun, and grabbing brass rings like his Eagle Scout badge and the graduate school fellowship. His failures after leaving graduate school had led him, by late 2010, to a crisis of confidence in which he turned away from his friends and broke up with his girlfriend for a spell.

"I felt ashamed of where my life was," he wrote in the digital diary he kept on his laptop. "More and more my emotions and thoughts were ruling my life and my word was losing power."

Silk Road was, in some sense, a last heave—a Hail Mary in the parlance of Ross's football-mad hometown. By the time he got it open in late January, he had, by his own accounting, gone through $20,000 of the $30,000 he had to his name.

When Silk Road finally opened up to anyone with a Tor web browser it was a simple site, with pictures of Ross's mushrooms next to their price in Bitcoin. At the top, there was a man in a turban riding a green camel, which would come to be the site's trademark image. Within days, a few people signed up, and the first orders came in for Ross's mushrooms. Soon thereafter, the first vendors joined in, offering to sell their own illegal wares. By the end of February, twenty-eight transactions had been made for products including LSD, mescaline, and ecstasy. Ross's growing confidence was evident from a message he posted on the Bitcoin forum from his new screen name: silkroad.

"The general mood of this community is that we are up to something big, something that can really shake things up. Bitcoin and Tor are revolutionary and sites like Silk Road are just the beginning," he wrote on the forum.

In his own diary, Ross was more frank: "I am creating a year of prosperity and power beyond what I have ever experienced before."

●

CHAPTER 7

March 16, 2011

The response to Silk Road on the Bitcoin forums was initially somewhat tepid—only a few people chimed in. But it got much more attention on the most widely used message board for hackers—4chan—and new Silk Road members were soon pouring in, along with orders. By mid-March, the site had over 150 members. That was, in fact, more than Ross was equipped to handle. He had to return again and again to the friend who had been helping him with the code, to figure out how to deal with all the traffic. When the site went down on March 15, he chatted his friend Richard Bates in a panic.

"i'm so stressed! i gotta get this site up tonight," Ross wrote.

"I'm not sure how this stuff works," Richard wrote back.

"i wish i did," Ross responded.

One of the people who visited the site while it was temporarily offline was the host of a popular libertarian radio program in New Hampshire, *Free Talk Live*, who was broadcasting live at the time. Ian Freeman and his cohost had been introduced to Bitcoin earlier in the year by Gavin Andresen, a regular listener who thought the show could reach an audience that would be sympathetic to Bitcoin. At a lunch with Gavin, the hosts of *Free Talk Live* had shown interest, but ultimately went away unconvinced. Who was going to have an incentive to use this? they asked. Their views, though, changed dramatically less than two months later when they learned about Silk Road.

"All of the sudden my interest has been piqued," Freeman said on the air.

Freeman and his cohosts did their best to explain how Bitcoin and Silk Road worked and they debated the possibility that Silk Road was a trap set up by the CIA. But the hosts agreed that Silk Road was something utterly new, harnessing Bitcoin to enable a type of transaction that was, for all intents and purposes, not possible before—an online drug purchase. What's more, getting cocaine or LSD delivered to your home—or a rented mailbox—seemed highly preferable to meeting a sketchy dealer at some dark rendezvous.

When Freeman tried to get on Silk Road while he was on the air, and found it was down, he wondered if it had all been a mirage. But when he had been on the site shortly before, he had seen 151 registered users and 38 listings. Someone had recently delivered ecstasy tablets from Europe to the United States, taped to the inside of a birthday card. Here was something that could take advantage of Bitcoin's unique qualities and help it grow.

"This could be the killer application for Bitcoin," Freeman said.

When Ross learned about the broadcast a day later, he had gotten Silk Road up again, and he wrote to his friend Richard Bates with a mixture of fear and pride.

"my site had a 40 minute spot on a national radio program," Ross wrote in a chat session with Richard.

"friggin crazy, you gotta keep my secret buddy," Ross added.

"I haven't told anyone and I don't intend to," Richard wrote back.

"i know i can trust you," Ross responded.

•

ONE OF THE many listeners who heard the conversation about Silk Road on *Free Talk Live* was Roger Ver, an American entrepreneur living in Tokyo, just a few miles from Mark Karpeles.

In comparison with many Bitcoin aficionados, Roger had a rather happy upbringing in the Bay Area, where he grew up with one sister and two half brothers. He had been a natural at the strategy game *Magic: The Gathering*—so good that he traveled on an amateur circuit to play competitively. But he was also on a wrestling team, and he and his brother both spent many afternoons fine-tuning their muscle cars—Roger's, a Mercury Capri; his brother's, a Mustang.

At the age of twenty, Roger signed up to run for the California state assembly as a libertarian candidate, vowing never to take a government salary. In the midst of his campaign for the assembly, federal agents arrested Roger for peddling Pest Control Report 2000—a mix between a firecracker and a pest repellent—on eBay. Roger had bought the product himself through the mail and he and his lawyer became convinced that the government was targeting Roger because of remarks he had made at a political rally, where he had called federal agents murderers. He would be the only person arrested for selling Pest Control Report 2000 through the mail and the prosecutors showed no leniency. Hit with felony charges, he was sentenced to ten months in prison after agreeing to plead guilty.

The experience turned Roger's libertarian ideas from a political cause to a personal crusade—he believed the government was out to get him. In prison, Roger taught himself Japanese, and the day his probation was up he flew to Japan to start a new life, free from the United States government. Japan's orderliness appealed to him. That and he had a thing for Japanese women.

It was during a brief trip back to California to see his family that Roger sat down to breakfast listening to a month-old *Free*

Talk Live podcast on his iPod. When the hosts started talking about Bitcoin, something snagged in his mind and he stopped what he was doing. Many Bitcoin fanatics would later talk about their ecstatic moments of conversion to the Bitcoin cause, but few were as extreme as Roger's. While the podcast was still playing, Roger did a search for Bitcoin on the laptop he had on his kitchen table and began making his way through everything he could find.

He was so entranced by the idea of a financial system outside the control of the government that he read clear through the night to the next day. After a short nap, he began reading again and went on reading for a few days until he eventually felt so weak, and so gripped by a sickness taking over his throat, that he called a friend and asked to be taken to the hospital. There he was connected to an IV sack that pumped antibiotics and sedatives into him. It might have been the drugs, but as he lay in his hospital bed, he felt he had found a kind of promised land that he had been waiting for all of his short life—the Galt's Gulch he had been searching for like a libertarian Indiana Jones.

Roger had an intuitive sense of the way markets worked long before he had developed his market-centric ideology. When Roger was in fifth grade, he cornered the market on Lindy dollars, a school-wide currency named for a beloved teacher, after realizing that a Lindy dollar was not worth the same as a real dollar, as most students assumed. Using his Lindy dollars, Roger bought up all the Rice Krispies treats and brownies at the school bake sale and once there were no other sellers, jacked up their prices. The other students quickly paid Roger's prices, realizing they had no other use for their Lindy dollars.

Roger launched a business, Memory Dealers, during his first year at De Anza College in Cupertino, just after the tech bubble burst, when bankrupt companies began selling their computer hardware cheap. He scooped up all the hardware he could find and sold it online. The business became so successful that he dropped out of school after his first year. By the time he discovered Bitcoin, his company had thirty employees and sales of around $10 million a year, which paid for Roger's Lamborghini Gallardo and his luxury apartment in Tokyo, just a few blocks from the flashing, teeming transit hub and commercial district of Shibuya.

In April 2011, after hearing about Bitcoin on *Free Talk Live*, he used his fortune to dive into Bitcoin with a savage ferocity. He sent a $25,000 wire to the Mt. Gox bank account in New York—one Jed had set up—to begin buying Bitcoins. Over the next three days, Roger's purchases dominated the markets and helped push the price of a single coin up nearly 75 percent, from $1.89 to $3.30.

At the same time that he was buying, Roger announced on the Bitcoin forums that his computer hardware company, Memory Dealers, would immediately begin accepting payment in Bitcoin. Not long after that, he turned a regular Memory Dealers' advertisement that he paid for on *Free Talk Live* into an advertisement for Bitcoin and crowdsourced the copy for the ad from the Bitcoin forums. Soon enough, he had put up a gold-and-black billboard, on the side of an expressway in Silicon Valley, with an enormous Bitcoin emblem and the phrase "We Accept Bitcoin," over the Memory Dealers web address. The crowd on the forums went wild.

"God I love Bitcoin!" one user wrote.

"We needed this," another said.

Roger said he was looking to do even more: "I promise I'm doing whatever I can to help make Bitcoin succeed (Billboards, National radio ads, etc.)."

Roger's appearance on the scene coincided with the first mainstream news coverage for Bitcoin, which helped push the price up, and, in turn, led to more mainstream news coverage. In the first such article, on *Time* magazine's website, Jerry Brito, a fellow at the libertarian-oriented Mercatus Center at George Mason University, was given space to discuss why Bitcoin might matter:

> Law-abiding citizens can carry on their affairs without
> anyone snooping on them or telling them what they can
> and can't do. Want to contribute to WikiLeaks or some
> other politically unpopular organization? No problem.
> Live under a repressive regime and want to buy a repressed
> book or movie? Here's how. No wonder the Electronic
> Frontier Foundation calls Bitcoin "a censorship-resistant
> digital currency."

A few days later *Forbes* magazine did its own lengthy and positive story on Bitcoin, noting that the virtual currency "cuts across international boundaries, can be stored on your hard drive instead of in a bank, and—perhaps most importantly to many of Bitcoin's users—isn't subject to the inflationary whim of whatever Federal Reserve chief decides to print more money."

Until very recently, Bitcoin had been kept alive almost entirely by computer programmers who played around with the Bitcoin software themselves. Now it was attracting a new breed of participant, like Roger Ver, who could not understand the code, but for whom the political possibilities behind Bitcoin were enough of a draw.

SATOSHI NAKAMOTO PICKED this moment to finally disappear for good. The author of the Bitcoin software hadn't posted to the forums since December, but he had continued to e-mail with a select number of developers, including Gavin, Martti, and Mike Hearn, a Google programmer in Switzerland, who got drawn into the project after the WikiLeaks blockade. In late April Hearn politely asked how involved Satoshi intended to be moving forward.

"Are you planning on rejoining the community at some point (e.g. for code reviews), or is your plan to permanently step back from the limelight?" he asked.

"I've moved on to other things," Satoshi wrote back. "It's in good hands with Gavin and everyone."

A few days later, Satoshi wrote a slightly peeved e-mail to Gavin about an interview he had recently given to another online radio show.

"I wish you wouldn't keep talking about me as a mysterious shadowy figure," Satoshi wrote. "The press just turns that into a pirate currency angle."

Gavin wrote back acknowledging the point. He also told Satoshi that he had received from the CIA an invitation to speak about Bitcoin, which he was planning to accept.

"I hope that by talking directly to them and, more importantly, listening to their questions/concerns, they will think of Bitcoin the way I do—as a just-plain-better, more efficient, less-subject-to-political-whims money," he wrote.

Gavin never got a response and assumed that Satoshi had been turned off by the idea of Bitcoin fraternizing with the most intrusive arm of the American government.

Satoshi's final e-mails went to Martti, whom Satoshi asked to take full ownership of the Bitcoin.org website.

"I've moved on to other things and probably won't be around in the future," Satoshi wrote to Martti, in early May, before transferring the site to Martti and disappearing into the ether.

Martti took responsibility for the site, but he had otherwise almost entirely stopped his work on Bitcoin. With the price rising, he sold more than half of his twenty thousand or so Bitcoins and bought himself a nice apartment in Helsinki. Both Martti and Satoshi seemed to recognize that the community had grown large enough that it no longer needed either of them.

•

THIS WAS THE moment that many early adopters had been waiting for. Bitcoin was getting mainstream attention and being taken
seriously by important people. By mid-May, the price of a single
Bitcoin was approaching $10.

Thanks to Silk Road, Bitcoin was being regularly used for the
first time as a medium of exchange for real, if illegal, things. This
was not enough to allow Bitcoin to claim the mantle of money,
which had several properties that Bitcoin lacked. But Bitcoin could
now meet some definitions of a currency, a label that had been
purely aspirational through 2009 and 2010.

"My wife isn't calling it a 'pretend money project' anymore,"
Gavin told the others gathered on the Bitcoin chat channel one
morning.

But Gavin didn't let this go to his head. He avoided the urge
to buy Bitcoins and speculate on their rising price, as everyone
else seemed to be doing. He had promised his wife that while he
would spend his time on the project, he would never spend any of
the family's money. At this point, it was also evident to Gavin that
the price and power of Bitcoin were no longer reliant just on the
strength of the underlying Bitcoin protocol. People moving into
and out of the virtual currency were using services that people
had built on top of the protocol, and it was quickly becoming
clear that these services were not equipped to deal with the rapid
growth.

In Tokyo Mark Karpeles had to rush home from his hon-
eymoon with his new Japanese wife—whom he had met a few
months earlier, not at a bar, but in the office building where he was
working—to try to fend off hackers who had launched a denial-
of-service attack on Mt. Gox. The attackers said they would relent
only if Mark paid a $5,000 ransom.

"This was—of course—denied," Mark explained to his users. "We do not negociate with internet terrorists!"

But it took days for Mark to install the necessary protections against what was a fairly standard attack.

In Texas, Ross had shut down his used book business so that he could work on Silk Road full-time. He was staying up late, furiously trying to rewrite his site from scratch so it would be able to withstand both the traffic and the hackers who were already targeting him. Silk Road now had over a thousand people registered, ten times more than it had just two months earlier. In mid-May, to get the new version online, Ross had to shut the site down for a few days, which turned into one of the more stressful periods he had endured.

"Updating a live site to a whole new version is no easy task," he wrote in his diary. "You don't realize how many little pieces lay on top of one another so it works just right (at least when you code poorly like my amateur ass was doing). So for about 48 hours it was stop and start on the switch, but I finally got there and it was working."

While Silk Road was down, the price of Bitcoin entered a short period of decline, suggesting just how important the site was for the fate of the virtual currency at this point. Silk Road users showed up on the Bitcoin chat channel asking if there was anywhere else they could score some drugs. When Silk Road came back online, the price of Bitcoin picked up again.

But the real onslaught began on June 1 when the gossip/news website Gawker published an in-depth story about Silk Road, based on interviews with people who had purchased and received LSD and purple haze pot from the site. There were now 340 different items available, including tar heroin and Afghani hash.

In the days immediately after this story came online, over a thousand new people were registering for Silk Road every day and the price of a Bitcoin on Mt. Gox shot up, crossing $10 for the first time the day after the Gawker story and $15 two days later.

The growth of the black market was something many of the old Cypherpunks had wanted to enable by creating an anonymous currency—in the 1990s some of the Cypherpunks had even talked about a "Digital Silk Road." But now that it was actually here, it was causing much more mixed feelings in the Bitcoin community. While Martti had welcomed the site and Roger Ver looked on approvingly, many of the Bitcoiners who were more interested in technology than politics thought this was the worst thing that could happen to the Bitcoin network. Gavin tried to personally distance himself and Jeff Garzik, a programmer living in North Carolina who had become one of the steadiest contributors to the Bitcoin software, wrote to Gawker to explain that Bitcoin was actually less anonymous than most people believed, owing to the record of all transactions on the blockchain. Sure, the blockchain didn't have names, but Garzik explained that the police would probably be able to determine the identity of users through sophisticated network analysis.

"Attempting major illicit transactions with Bitcoin, given existing statistical analysis techniques deployed in the field by law enforcement, is pretty damned dumb. :)," Garzik wrote.

In conversations with other developers, Garzik was less worried about Silk Road users getting caught and more concerned about all the negative attention that Silk Road would bring if it continued to grow. The worst fears of people like Garzik were borne out on June 5 when Senator Chuck Schumer of New York held a heavily covered news conference, at which he decried the brazen business of Silk Road and called for prosecutors to shut it down. He described Bitcoin as an "online form of money laundering used to disguise the source of money, and to disguise who's both selling and buying the drug."

Rather than scaring people away, Schumer's commentary—and the deluge of media attention it received—brought on yet another surge of interest, sending the price of Bitcoin on an Icarus-like rise that had it at $30 within two days. That was a 600 percent rise from a month earlier, and a 9,000 percent increase from six months earlier. Silk Road now had ten thousand members.

Ross had, by now, fully recouped his initial investment—earning $17,000 from the sale of his mushrooms, and $14,000 from commissions collected on the sales made by others. But the news out of Washington strained Ross's already frayed nerves.

"I was mentally taxed, and now I felt extremely vulnerable and scared," he wrote in his journal. "The US govt, my main enemy was aware of me and some of its members were calling for my destruction. This is the biggest force wielding organization on the planet."

When Ross shut the site down in mid-June, to take a breather, he wrote on the Bitcoin forums that his little experiment had claimed way too much attention: "We'll do our best to get out of the spotlight and hopefully the merits of Bitcoin will become the focus."

But for regular Bitcoin companies, the situation wasn't going much more smoothly. Around the same time Silk Road went down, Mark Karpeles found himself unable to process withdrawals from Mt. Gox for four days. The problems helped pull the price of Bitcoin down almost as quickly as it had gone up. But even as the price settled down, below $20, something in the air was different. Some of Bitcoin's youthful innocence seemed to be gone.

Just a few months earlier—and even a few weeks earlier—the forums and chat channels had felt like a cozy global community. All the main characters could be found online talking to each other at almost any hour.

Now, everyone was too busy to chat, or was put off by all the negative energy. Mt. Gox users were on the forums complaining about Mark's silence as his exchange struggled and trades got delayed. In the chat rooms, a few upstart exchanges that were attempting to challenge Mt. Gox slammed Mark and his maintenance of Mt. Gox. There were a growing number of signs that Mark was indeed falling behind. In May he had hurriedly decided to move Mt. Gox into an expensive office tower, but so far he had been able to find only one employee who was willing to take the risks involved in working on Bitcoin. Jed McCaleb sent Mark suggestions for how to improve the site but Mark never responded.

Much of the tension in the broader Bitcoin community seemed to be a result of the deluge of curiosity seekers and pranksters, who overwhelmed the chat channel with inane commentary. In June, over 15,000 new people joined the forums, more than doubling the membership and leading to 152,000 new postings.

Bitcoin was supposed to be a new kind of community with no central authority, powered by the people who joined it. That had worked until now because the people involved wanted to see it succeed. But what if the people joining in had no such interest? Should some authority figure intervene and, if so, who could it be?

Some of the leading developers working with Gavin suggested that moderators should more aggressively police the forums and potentially even move the forums from Bitcoin.org, so that the conversations on the forums didn't look as though they had some official status within Bitcoin.

Martti, who had been given final say over the websites by Satoshi, was uneasy about these changes. He said he had long avoided determining what should and should not be discussed on the forum, as long as illegal transactions weren't happening on the forum itself.

Gavin largely stayed out of the public debate—he knew it wasn't worth fighting—but he quietly found a way to move forward by creating a mailing list dedicated to Bitcoin development that would be easier to control, a move that did not go over well with everyone.

Around the same time, Gavin made his visit to the CIA to present Bitcoin to a conference on emerging technology. He reported back immediately to the forums and was transparent about what he had said during his visit and what the response had been (everyone at the CIA meeting seemed to be interested). Many people on the forums were supportive of his decision to make the visit, but not everyone was. Those debates, though, were quickly overshadowed by bigger questions about whether the people building this community had the skills to keep it growing.

●

CHAPTER 8

June 19, 2011

The Tokyo sky outside Mark Karpeles's window was still dark when the iPhone on his bedside table jolted him awake just after 3 a.m. Mark was still trying to get his bearings when he picked up the phone. On the other end was the panicked voice of his friend William, a Frenchman living in Peru who had first introduced Mark to Bitcoin back in 2010.

For the last few weeks, William had been helping Mark keep up with the seemingly irrepressible expansion of Mt. Gox, which had grown from three thousand users in March to over sixty thousand users in June. Just how little Mark was prepared for the recent growth was clear from what William was trying to tell him on the phone. Something about the exchange's servers slowing down to a glacial pace—and the price of Bitcoin plummeting from $17 to 1 penny in less than an hour.

Suddenly alert, Mark leaped out of the bed he shared with his new wife and ran to the home office in their compact Tokyo apartment, one floor up from the narrow street. Mark was not generally known for moving fast—most who met him immediately noticed his slothlike way. But once he had his Mt. Gox administrative account up on the screen, Mark wasted no time in bringing the crisis to a screeching halt. He shut down the link between the Mt. Gox website and his server and moved Mt. Gox's 432,000 Bitcoins—some $7 million at yesterday's prices—to a new address that had a more secure password.

These moves were enough to stem the run on Mt. Gox, but immense damage had already been done. Hackers had enjoyed nearly an hour to do their work, while confused and terrified Bitcoin users looked on. Starting at around 2:15 in the morning in Japan, the hackers had begun selling large quantities of Bitcoins, pushing the price down dramatically.

"Everyone! Panic sell!" someone wrote on the chat channel, seeing the price dive.

"Holy fucking sht," another wrote.

One user had the presence of mind to record the charts showing the decline and narrate a video of it in real time. Others, who had dollars in their Mt. Gox account, saw an opportunity and began buying up the cheap Bitcoins. The selling continued until 260,000 Bitcoins were purchased for $2,600 shortly before 3 a.m. Japan time—a 99.94 percent discount from their value just an hour earlier.

After Mark had shut everything down, he sat in his dark apartment and began to piece together what had happened. The user logs showed that someone had signed in with the administrator account of Jed McCaleb, the Mt. Gox founder who was still helping Mark out. The computer appeared to be in Hong Kong, but it was likely the hacker was porting in to a computer there from elsewhere. The Mt. Gox software enabled the hacker to change the balances in accounts and he created over 100,000 new Bitcoins out of thin air and put them in a new Mt. Gox account. These were not real coins on the official blockchain; they existed only in Mark's accounting system. But that was enough for the hacker to begin using them on the Mt. Gox exchange.

The hacker had clearly planned in advance and knew that Mt. Gox allowed users to withdraw only $1,000 worth of Bitcoins at a time. In order to maximize the amount of Bitcoins that could be withdrawn, the hacker began selling some of the newly created coins to push down the price. As the price dropped, it was possible to withdraw more and more Bitcoins under the $1,000 limit, until the relatively primitive design of Mt. Gox came to its rescue. As the servers slowed to a crawl, owing to the traffic created by the hacker, withdrawals suddenly became impossible. By the time Mark got up, most of the hacker's Bitcoins were still stranded inside Mt. Gox, though hundreds of thousands of coins had already been sold at distorted prices.

It was not until an hour after he first got online—and two hours after the melee began—that Mark posted any kind of explanation to the Bitcoin forums. At that point, he gave the basics of what he knew and said that the site would be down indefinitely. He also announced his intention to cancel all the trades made with the Bitcoins created by the hacker, a move that drew an immediate backlash from buyers, who believed that they had gotten thousands of those Bitcoins on the cheap. Although many expressed anger that Mark was violating one of the fundamental tenets of Bitcoin—the irreversibility of Bitcoin transactions—Mark could do so because trades on Mt. Gox happened only within the company's system, not on the actual blockchain (Mt. Gox interacted with the blockchain only when coins moved into and out of the company).

The scope of the questions soon expanded, especially after it emerged that the hacker had stolen a copy of Mt. Gox's customer database, with everyone's e-mail addresses, and posted it on the Internet. There was bewilderment that Mt. Gox administrators had needed only a single password to log in, not the multiple passwords that most financial websites required. And Mark's system had not checked on the IP address and location of users to look for abnormal activity.

"Frankly, we are fortunate that our hackers have been stupid and lazy so far," Jeff Garzik, the North Carolina programmer, said to some other developers.

On top of these programming mistakes, the released customer database demonstrated how few measures Mark had taken to stay compliant with international rules designed to stop money laundering. Mark had just e-mail addresses for most of his users, much less than financial regulators generally expected. Of course, it wasn't clear what regulations Bitcoin would fall under, if any. But there was now real money flowing into and out of Mt. Gox, making the exchange an easy target for government prosecutors if they decided to look.

THE FIRST SIGN of any relief for Mark came in an e-mail that popped into his inbox later that morning.

Hey Mark—

If you guys need any physical help, I'm available. I can be at your office within 10 minutes.

I'm not sure what I can do to help, but I can help with phones or emails or anything you need for a day or two until you get things calmed down.

The e-mail came from Roger Ver. From Roger's glass-walled sixteenth-floor apartment, in one of Tokyo's most exclusive residential towers, he could see the Cerulean Tower, where Mark had recently set up Mt. Gox's offices. Since discovering Bitcoin in April on *Free Talk Live*, Roger had dedicated many of his waking hours to thinking up new ways to promote the technology. In a conversation right before the crash he had said something that would become a standard line for him: "Bitcoins are the most important invention since the internet itself. They will change the way the entire world does business."

At this point, though, Roger knew that Bitcoin relied as much on Mt. Gox's survival as on the Bitcoin protocol itself, and he wanted to make sure that Mt. Gox would survive so that Bitcoin could as well.

By the time Roger sent his e-mail, Mark had driven in his souped-up 2009 Honda Civic from his apartment to his new office. Mark quickly connected with Roger on Internet chat—Mark's preferred method of communication—and asked him to come right over. He needed people who could speak English and sort through the thousands of incoming e-mails from confused customers.

When Roger showed up at the bare-walled office, he was an even more forceful and impressive presence than he seemed online. He had the lean, muscular physique of a wrestler, which is what he had once been, and the buzz cut and big smile of a politician, which is what he had once wanted to be. What's more, he came with his Japanese fiancée, Ayaka, and one of his employees from Memory Dealers, whom he put at Mark's service.

Roger, on the other hand, had to adjust his judgments of Mark in the other direction. Mark had the chubby look of a big child and the nervous crooked smile of someone who was not entirely comfortable with direct human contact. His wardrobe was heavily reliant on T-shirts with puns about programming languages. His heavily accented English made him difficult to understand. Mark's only staff member was a young Canadian with no programming expertise who had been hired a few weeks earlier. Roger put all this aside for the time being and dived into the flood of customer-support requests.

Roger brought an energy unlike anything that Mark had seen before. As he plowed through complaints and requests, Roger also managed to convince an old friend to get on a flight from California to help the Mt. Gox rescue effort.

Roger and the friend who came to Tokyo the next day, Jesse Powell, were a somewhat unlikely pair. In contrast to Roger's clean-cut, buttoned-down appearance, Jesse had long blond hair and had used money from his startup to found an art gallery in his hometown, Sacramento. But Jesse and Roger had met when they were teenagers and both playing the card game *Magic* competitively. The strategy game appealed to both young men—and many of the other youngsters who later found Bitcoin—because they liked the idea of finding unexpected solutions to complex problems. Later on, the same instinct had led both of them to the martial art jujitsu. A mixture of Japanese and Brazilian influences, jujitsu gained renown as a way for smaller, less muscular people to disarm and defeat larger opponents. Libertarianism and Bitcoin were alluring to Roger and Jesse for much the same reason, owing to the deceptively simple answers they promised for much bigger problems.

Roger had chosen his apartment in Tokyo largely because it was near his jujitsu studio, or dojo, and during Jesse's visit to help at Mt. Gox, the men went to the dojo to grapple with each other and let off steam. But they spent almost all of their time working through the constantly growing pile of e-mails that had been sent to info@mtgox.com.

Mark, for his part, spent these days silently parked in front of his computer, investigating the cause of the hack. He determined that the attacker had gained access to Jed's Mt. Gox administrative account by either guessing the password with the brute force of a computer program or by gaming the system that allowed users to create new passwords. In the end, Mark calculated that the site had lost only a few thousand Bitcoins, which he promised to reimburse with the company's money.

Mark then moved on to rewriting the Mt. Gox code so that he could reopen the site. Two days after the crash, he appeared briefly, via Skype, on *The Bitcoin Show*, a relatively new online production created by an enthusiast in New York. Mark took the opportunity to blame the code he inherited from Jed McCaleb, which he said had "a lot of problems."

"The new system was written from scratch with absolutely no code from the old system," he said. "It was made from state of the art techniques."

Two days after that, Mark made a transfer of 424,424 Bitcoins that was visible on the public blockchain, in order to prove that he had his customers' coins.

"Ready guys?" he asked, right before making the move. "Don't come after me claiming we have no coins after that."

"Hopefully I'll be able to work without getting too much disturbed after that," he said.

Roger and Jesse were initially impressed by Mark's calm during the crisis. Every day he sat quietly at his desk, eyes fixed on the screen. But as the week progressed, Mark's silence put him at an uneasy distance from the surrounding world. Jesse and Roger grew concerned that all Mt. Gox's technological and financial affairs were in the hands of one person, with no one else in a position to question his decisions or stand ready if things went wrong. They also worried about Mark's ability to prioritize tasks properly. They frequently noticed that when Mark was supposed to be working on fixing the site, he was instead on the Mt. Gox chat channel, trying to address customer complaints. At the end of the week, Roger and Jesse asked what time they should come in the next day.

"Oh no," Mark said. "We can just start again on Monday."

"But this site isn't even back up," Roger said. "I think we should keep working until we get it up."

Mark said something about the office tower being closed during the weekends and shut off further conversation. While walking back to Roger's apartment, Roger and Jesse wondered at Mark's lack of urgency.

Mark himself worked through the weekend, from his apartment, and opened the site for trading on Monday morning. As soon as this happened, the price of Bitcoins began falling. In the week that Mt. Gox had been closed, the public perception of Bitcoin had taken a decided turn for the worse, with a series of news articles suggesting that the hack marked the likely end of Bitcoin. The day after Mt. Gox reopened, *Forbes*, which had been among the first to write positively about Bitcoin, said that "it's likely to go the way of other online currencies," the first of many public obituaries for Bitcoin.

CHAPTER 9

●

July 2011

In the weeks after Mt. Gox got back online, it was contending with new exchanges that had been started during the busy spring. But for the people who stuck around Bitcoin after the Mt. Gox attack, there was seemingly no end to the bad news.

In July, the founder of a small Polish Bitcoin exchange, Bitomat, announced that he had accidentally deleted the files where he kept the private keys to the Bitcoin addresses at which his customers' 17,000 Bitcoins were stored. The coins were still visible on the blockchain, but without the private keys, nothing could be done with the coins.

This pointed to a danger that was the flip side of one of Bitcoin's supposed strengths. Satoshi Nakamoto had designed Bitcoin so that each user had complete control over the coins in his or her addresses. Because only the person with the private keys to an address could access the coins assigned to that address, governments could never seize the coins and banks weren't needed to hold them. This design also meant that the coins themselves weren't stored on any particular computer; if a computer holding a wallet file with the private keys crashed, the coins were still on the blockchain, as long as the owner still had copies of the private keys.

But the design also meant that if a person lost the private keys for a particular address and had no backup, there was nothing anyone could do to access the coins held by that address. People were already taking precautions to guard against this, writing down the private keys on a piece of paper or maintaining backups. But what if the piece of paper was lost, or if the secure document with the keys in the cloud, as in Bitomat's case, was accidentally deleted, along with its backups? Not everyone, it turned out, was good at keeping track of valuable things.

Another incident just days after the Bitomat losses reminded everyone that the companies holding customer Bitcoins had another vulnerability—the integrity of the people running the companies. The losses this time happened to customers of MyBitcoin. The site, which had been around for over a year, provided a simple online wallet and held the private keys for all of its customers, so the customers didn't have to worry about losing keys.

In late July coins started mysteriously disappearing from MyBitcoin wallets. The founder of the site, a man who called himself Tom Williams, was unresponsive and soon enough all the wallets were frozen. Customers realized that they had no idea who Tom Williams actually was. On the forums, a group of users formed a vigilante online posse to try to hunt down Williams, but after making initial progress they lost the trail. It quickly became clear that Tom Williams, whoever he was, had now disappeared with everyone's Bitcoins and there was nothing anyone could do to get them back. In the days after his disappearance, the price of a Bitcoin fell to $6.

•

THE SCANDALS AND steadily declining price of Bitcoin over the summer of 2011 drove away most of the crowds that had been drawn in when the price was shooting up a few months earlier. The future for Bitcoin, a technology that relied on maintaining the trust of its users, seemed about as bleak as it had ever been.

But the disappearing crowds were a bit like a receding tide. They exposed what had been left behind and it was not an altogether disheartening scene. Yes, there were fewer people, but most of the serious programmers who had gotten involved in Bitcoin earlier on had stuck around.

For people like Gavin Andresen and Jeff Garzik, the problems at Mt. Gox and MyBitcoin were evidence for why a decentralized financial network like Bitcoin was needed. Both Mt. Gox and MyBitcoin were centralized companies and they failed because of the amount of power and money that had been placed in the hands of their operators. With Mt. Gox, the hacker had needed to get only one password to access the entire system. And because Mark kept tight control over all the code for Mt. Gox, his customers couldn't review the software and chip in with suggestions and improvements of the sort that could have helped avoid the hack. The Bitcoin protocol, on the other hand, had been slowly improved over time by all the people looking at it, and had continued working as intended throughout the various crises.

As the summer went on, it was evident that Bitcoin had not just kept its hold on the experienced programmers—all the excitement in June had actually drawn the attention of many new programmers, who understood the distinction between the Bitcoin protocol and the current crop of Bitcoin companies.

Mike Hearn, the British engineer working in Google's Swiss

offices, had created an e-mail list for Google employees interested in Bitcoin, and through the summer of 2011 it had grown to over a hundred people. On the list, Google employees conversed about the new ideas and potential that were contained within the Bitcoin protocol.

One Google engineer in the company's Mountain View headquarters, Charlie Lee, sent Hearn a check for $3,000 by PayPal in exchange for a batch of coins. At the same time, Lee wrote to his family with twelve bullet points with reasons for giving it a look, including:

- The whole system is distributed and decentralized. It's a peer to peer system. No government can shut it down even if Bitcoins were outlawed.
- The system is self sustaining. The miners (i.e. p2p nodes) have incentives to keep mining, which helps secure the whole system. The more the system is secure, the more the users will trust in Bitcoins and use them. And the more people use them, there's more incentives for the miners.
- Everything is defined by its source code and it's opened source.

Five or six other Google employees began developing new Bitcoin software to make the network easier to access. Mike and the other Googlers were taking advantage of the company's policy of allowing its employees to spend 20 percent of their working time on non-Google experiments. Mike used this time to develop BitcoinJ, a codebase that made it possible to work Bitcoin into mobile phone applications. This was a significant step for the virtual currency. Before this, everyone who wanted to use the system had to download the Bitcoin software and a copy of the entire blockchain. That was, by now, a large file, and its size made it all but impossible to use Bitcoin on a phone or anywhere other than a home computer. Mike was making it possible for people to use Bitcoin without actively participating in the network, something that would open it up to new audiences with less technical expertise.

The work caused some disquiet among Mike's superiors at Google, who feared that his work could earn Google unwanted scrutiny if the government decided it didn't like Bitcoin. But he fought to keep working on it, and won. And not all the higher-ups were so cold to the idea.

The head of Google's payments division, Osama Bedier, called Mike in to get a tutorial on the technology. Mike knew that Google had long struggled with how to build its own digital payments system. The program that Abedier was working on, known as Google Wallet, was not creating a new payment system—instead it was looking to provide a new means of using existing credit cards and bank accounts online. All the fees and restrictions with credit cards and bank accounts still applied to Google Wallet.

Mike gave Abedier a lesson on the basics of a virtual currency that had no central authority and essentially no transaction fees. When Mike finished his presentation, Abedier told him, "I would never admit it outside this room, but this is how payments probably should work."

The Bitcoin developers who were not at Google generally continued their work with no compensation at all. For Gavin, who had become the lead programmer for the Bitcoin protocol, the work had become a full-time but unpaid job. He was working out of the little office he shared with his wife in their Massachusetts home. His desk chair was next to an old radiator, which rattled in the winter, and a window air-conditioning unit, which rattled in the summer.

The passion that Mike and Gavin had for Bitcoin had little to do with where the technology stood in the summer of 2011. After all, it was still hard to actually buy much with Bitcoins. In August, when someone came up with a list of brick-and-mortar institutions that accepted Bitcoin, there were all of five entries. The programmers were also acutely aware of flaws in the Bitcoin software that would need to be fixed if the system were to grow.

But none of this distracted the programmers from their vision of what the Bitcoin software could do in the future. Some programmers were focused on the idea of micropayments, tiny online payments that are not possible with credit cards because of the minimum fees necessary for a credit card transaction.

Others were interested in the idea of immigrants sending money across international borders without using Western Union. Some imagined the sorts of smart contracts that Satoshi had described, which would allow people to sell a house without using expensive mortgage title companies and escrow services. Yet others had a more abstract idea of a future universal currency, as science fiction had promised.

IN ADDITION TO the coders, Bitcoin had kept its hold on many of the believers who were more interested in the ideals behind the virtual currency than the price. Over the summer, this crowd got a showcase on *The Bitcoin Show*, the web-only television show created by Bruce Wagner, a New Yorker whose enthusiasm compensated for his lack of experience producing television and his lack of knowledge about computer programming. Early in the summer, Wagner had begun planning for what he was calling the Bitcoin Conference & World Expo NYC 2011. He was not shy about his ambitions for the event, which he scheduled for late August:

I know for sure attendees are flying in from every

continent. Some on private jets.

This will be HUGE. No, definitely not just another Bitcoin meetup.

Major global press—tv, magazines, and newspapers, have confirmed that they will be here.

On the forums there were questions about whether anyone would show up. But the list of people promising to attend grew as the date approached.

Roger Ver flew to New York from Tokyo for the conference and shared a hotel room with Jesse Powell, who came in from Sacramento. Jed McCaleb flew up from Costa Rica. Mark Karpeles, consistent with his reputation, decided to stay in Tokyo, despite the fact that Mt. Gox was the major sponsor of the event. Charlie Lee, the Google engineer who had purchased $3,000 of Bitcoin from Mike Hearn, flew in from California. Gavin Andresen came down to New York in a MegaBus that left from a mall near his house in Massachusetts. Gavin was not the conference-going type, but the bus ticket was cheap and he couldn't resist the opportunity to meet all the people he had been interacting with online for the last year.

The conference was a rather apt representation of Bitcoin itself, with its odd mixture of chaos, community, snake oil, innovation, high-mindedness, and enthusiasm. While Wagner had initially suggested that the whole event would be held in the rather run-down OnlyOneTv studios, he ended up getting space at the Roosevelt Hotel in midtown Manhattan. The room was the smallest one on offer, a floor above the main conference center, with low foam-board ceilings. The handful of exhibitors, who had paid $130 to attend, were given card tables to set up their wares, just inside the narrow entrance to the room.

Wagner had promised three days of events, but in the end there were only three talks, taking up less than two hours, and they got started almost four hours late. Still, once everyone was in the room, there were almost a hundred people, and they buzzed with a childlike excitement at seeing these characters whom they'd known only as online avatars before. The event began with all those in the room introducing themselves, both by their screen name and by their actual name.

The first speaker was Gavin, who lived up to his folksy reputation. He recounted how he had learned about Bitcoin, and explained why he believed Satoshi had chosen to put him in charge.

"You can call me an idiot and yeah, whatever," he said, with a grin. "I know I'm not perfect so I tend not to rush into things rashly. Because I screw up quite regularly, my virtue is that I will listen to you if you tell me I'm screwing up."

He gave a wish list of things he wanted to work on—focusing on security and stability—and expressed his desire to see Bitcoin become "really boring" as it became more useful.

After two other, more technical speeches, the event was closed with a brief talk by Wagner, clad in a striped black dress shirt and a striped black sport coat. He seemed to twitch with eagerness.

"I'm just so so excited and honored to be here—to witness this. I love you all. It's just so freaking awesome. Right?" he said.

He had promised in the run-up to the event that he would "be making a HUGE HUGE HUGE announcement at the Conference. One you're all gonna be VERY excited about . . . when you hear it."

He built it up by first announcing that there would be another Bitcoin conference in New York in October 2012. Then he said there was going to be a Bitcoin conference in Amsterdam in June 2012. Finally he got to the conference he was planning in Pattaya, Thailand, only six months away.

"If that's not enough," he said, there would also be the first-ever Bitcoin cruise in Brazil.

The audience sat silent, with more than a few arched eyebrows, as if to ask—"Was that really it?" But Wagner did not pick up on the skepticism.

The crowd, though, had not come for Wagner. The attendees had come for each other. And as the brief planned portion of the conference concluded—after a big group picture—the conversation continued all evening and all night, moving to the Hudson Eatery, one of three restaurants that Wagner had convinced to take Bitcoin.

There, Roger Ver, the Tokyo entrepreneur, talked with the Google engineer, Charlie Lee, who described the computers that he had in his garage, mining Bitcoins. They were noisy, blowing out heat, and had begun to annoy Lee's wife. Roger offered to house the computers at Memory Dealers' offices in Silicon Valley.

Jesse Powell, Roger's friend who had helped out during the Mt. Gox crisis, found a kindred spirit in Mt. Gox's creator, Jed Mc-Caleb, who shared the same laid-back, nerd-cool sensibility. Jesse told Jed about his experiences over the summer in Tokyo with Mark Karpeles. And Jed told Jesse about his recent ideas for a new cryptocurrency that would not require Bitcoin's energy-intensive mining process. Meanwhile, Gavin was surrounded by people offering to help with the goals he'd set out in his talk. Despite his aversion to crowds, the event was intimate enough, and overflowed with enough enthusiasm that even he got into it.

The spirit in the restaurant was no small part of what was allowing Bitcoin to survive. A project that seemed aimed at furthering an even greater virtualization and atomization of our world was actually creating a sense of real-world community with people working together, animated by a shared sense of purpose for changing the world. The community, which mostly lived online, wasn't always this harmonious. But it was possible, and the sense of community was a significant draw for a group of people who didn't always find it easy to find like-minded people in the ordinary world.

When it came time to pay the bill, the waiter had little idea of how to actually handle the Bitcoins and it took over an hour to get everyone's money transferred. But no one much cared, or bothered to remark on the cumbersomeness of this supposedly space-age payment mechanism. It gave everyone more chance to talk.

CHAPTER 10

●

September 2011

When Roger Ver returned to Tokyo, he was immersed in plotting his next big Bitcoin campaign with a twenty-six-year-old who had marched up to him during the conference in New York and handed him a business card that read, "I am friends with Satoshi," under the name Erik Voorhees.

"We should talk," Erik had said to Roger.

With a confidence and poise that were notable for someone his age, Erik explained to Roger that since learning about Bitcoin from a Facebook posting just a few weeks after Roger came on the scene, he, Erik, had been intently watching Roger's work online, cheering him on from afar, and doing similar evangelizing for Bitcoin whenever he could.

Erik had recently moved back to the United States from Dubai, where he had gone for a job in real estate marketing after college. He and his college sweetheart had chosen to settle in a small seaside town in New Hampshire, where they joined the Free State Project, a movement founded on the idea that if several thousand ardent antigovernment activists gathered in one small state, they could influence the political direction of that state. New Hampshire was an obvious choice, with its motto, "Live free or die." *Free Talk Live*, the radio show that had introduced Roger to Bitcoin, was hosted by other members of the Free State Project.

Erik had grown up in the mountain town of Keystone, Colorado, where he had become an adept skier and mountain biker. In high school, he learned to DJ, playing house and techno music at local parties. As an undergraduate at the University of Puget Sound, he joined the Sigma Chi fraternity.

But he also had a more serious, political side that he got from his father, a passionate advocate for free markets and entrepreneurs who had built his own jewelry business. His father had been a competitive debater and urged Erik to follow in his footsteps, given Erik's smooth way with words. Erik, though, had discovered that he could not convincingly argue a point he did not believe in, and so he threw himself into advocating for the ideas he did believe in.

After the financial crisis, Erik became particularly fascinated by the role that central banks played in maintaining government power. He came to believe that it was only through printing money that governments were able to pay for their budgets and wars. Monetary policy had been one of the issues he was most passionate about when he joined the Free State Project. But when he discovered Bitcoin, he saw a shortcut to achieving his goal of a world without government power. Erik had largely abandoned his efforts to find a new job and went deeper into credit card debt so that he could spend his time evangelizing for Bitcoin.

"You don't have to try to vote your way into changing the world," he would tell anyone who listened. "If Bitcoin works, then it will change the entire world in a decade, without asking for anyone's permission."

Meeting Roger in person, Erik immediately detected that they shared more than just basic libertarian politics. They both occupied a more idealistic place on the libertarian spectrum, less interested in reducing taxes and more interested in stopping government-sponsored wars—looking up to the same thinkers who had motivated Ross Ulbricht. At the same time, neither Roger nor Erik was the type of anarchist-leaning libertarian who fought against authority figures and societal expectations of all kinds. Both men always looked presentable—usually clad in slacks and polo shirts—and generally approached conversation with a respectful and deferential tone.

At the conference, the two men had commiserated about the fact that even in the libertarian world, where Bitcoin should have had the easiest time winning fans, it had been slow going. Both of them had run up against lots of libertarians who doubted the American dollar, but did not see Bitcoin as a more stable or solid alternative.

The problem for many libertarians was their ingrained belief that money had to be backed by something with real value, like gold. One of the patron saints of gold bugs, the economist Carl Menger, had argued that all successful money arose from commodities that had some intrinsic value, even before they become money. From this perspective, Bitcoin appeared to have no chance—there was no independent demand for these virtual tokens on the blockchain. But Erik argued that it was the very virtual nature of Bitcoin that made it so valuable. Unlike gold, it could be easily and quickly transferred anywhere in the world, while still having the qualities of divisibility and verifiability that had made gold a successful currency for so many years.

By the time they left New York, Erik and Roger had hatched a plan to start winning over some of the libertarian doubters. Their goal was to get some actual Bitcoins into the hands of all of the fifteen thousand or so people in the Free State Project. Roger offered to donate the coins himself. It took some negotiations with the board of the Free State Project. Given its concern about privacy, the organization didn't want to hand over the e-mails of members. But Roger offered to send the board the coins so that it could send the coins out itself. To deliver the coins—0.01 Bitcoin for each person—Roger and Erik used a new program that Erik had been developing with a programmer he knew in Colorado.

Part of the goal was to show how Bitcoin could allow transactions that were not possible, or at best not easy, in the traditional financial system. Roger transferred his donation from Japan to New Hampshire without any fees or wait. Meanwhile, the size of the payments sent to each member was small enough that the fees involved in sending such a payment, using PayPal or a check, would have been greater than the payment itself. On top of that, the Free State Project could send the money to its members without needing any personal information—showing that this was, indeed, digital cash.

The whole thing was worked out by the beginning of October and, as part of the deal, the Free State Project began accepting donations in Bitcoin. The announcement from the Free State Project made the board members sound like converts: "Our eyes are on the long-term, the future, and Bitcoin is very exciting for our project and human freedom in general."

BITCOIN HAD THE good fortune of hitting hard times at a moment when there was a renewed willingness to rethink the foundations of the existing financial system.

On one side of the spectrum, the 2012 presidential campaign of Ron Paul was gaining steam in the fall of 2011, thanks in no small part to his discussion of the Federal Reserve and monetary policy. He argued that the central bank had encouraged the real estate bubble with low interest rates, and had done more damage by printing money after the crisis hit. Around the time that Erik was selling the Free State Project on Bitcoin, Paul likened the Fed's money printing to a drug addiction. He warned that if it wasn't reined in, the central bank would do itself in.

"The Federal Reserve will close themselves down eventually when they destroy money," Paul said on the campaign trail.

Meanwhile, a month after the Bitcoin conference, protesters took over Zuccotti Park in Manhattan and began what became known as Occupy Wall Street, taking aim at the government's decision to bail out the big banks but not the rest of the population. The Bitcoin forum was full of people talking about their experiences visiting Zuccotti Park and other Occupy encampments around the country to advertise the role that a decentralized currency could play in bringing down the banks. The people who had been attending the New York Bitcoin Meetup went to Zuccotti Park with flyers and cards offering an introduction to Bitcoin. Soon enough, a few branches of the Occupy movement began accepting Bitcoin donations. The anticorporate Occupy sentiment was even more widespread in the European Bitcoin community, where libertarianism had less of a foothold. An anarchist bar in a hip neighborhood of Berlin, Room 77, had been one of the first establishments to accept Bitcoin and it became a regular gathering spot for many of the European Bitcoin developers who were working with Gavin Andresen.

The different communities where Bitcoin was winning support were not always in agreement about what kind of future they were working toward. For many members of the Free State Project and the Ron Paul campaign, the problem was the excessive role of the government, which had created a subservient population that didn't know how to take care of itself. The Occupy Wall Street crowd was often OK with government, as long as it was serving the interests of the people, not of corporations and banks.

But in the wake of the financial crisis and the Iraq War, these people and movements generally shared a desire to take power and resources back from society's ruling institutions and return them to individuals. Both Occupy Wall Street and the Free State Project were ostensibly leaderless organizations that eschewed new power hierarchies.

Political scientist Mark Lilla has written about the onset, after the financial crisis, of a libertarian age, in which the shared values are "the sanctity of the individual, the priority of freedom, distrust of public authority, tolerance."

These principles, Lilla said, have been enough to bring together

> small-government fundamentalists on the American right, anarchists on the European and Latin American left, democratization prophets, civil liberties absolutists, human rights crusaders, neoliberal growth evangelists, rogue hackers, gun fanatics, porn manufacturers, and Chicago School economists the world over.

Few things occupied the common ground of this new political territory better than Bitcoin, which put power in the hands of the people using the technology, potentially obviating overpaid executives and meddling bureaucrats.

Not everyone in the Bitcoin world partook in the politicization of the technology, particularly among the developers. Gavin was generally sympathetic to libertarian ideas, but he also knew that some people did get lucky advantages in life—thanks to better educational systems and family situations—and it was these people with built-in advantages who tended to do best when government went away. He was also skeptical that political arguing did much to change people's beliefs. Jed McCaleb, meanwhile, openly chastised fellow Bitcoiners for their emphasis on the "libertarian, going to replace all other currencies, take over the world stuff."

"That just turns people off," he said. "The only important thing for people to know is that it is better than what people use now for online payments."

But the people ignoring Jed's advice ended up giving Bitcoin momentum at a time when it was otherwise lacking. Roger alone bought tens of thousands of coins in 2011, when the price was falling, single-handedly helping to keep the price above zero (and establishing the foundation for a future fortune). As Erik would joke, no one would be stupid enough to invest in a project as experimental as Bitcoin without some noneconomic motive for doing so.

"Who the hell is going to put their money into something so completely wacky?" Erik would say, with a self-disparagement that was somewhat unusual for such an ideological partisan. "You have to have an ulterior motive."

What's more, at a time when ideology was a major national talking point, the principles that were becoming attached to Bitcoin were helping it to win public attention, as a symbol of the new politics taking root in America.

THE IDEOLOGICAL UNDERPINNINGS of Bitcoin helped it win new followers, but the growing adoption of Bitcoin was also serving as a real-world test for these big ideas—and it didn't always bear out the hopeful assumptions of the followers.

Bitcoin had succeeded in its goal of giving its users control over their own money, without requiring a bank or any middleman to conduct transactions. But all the money that had piled up in Mt. Gox and MyBitcoin suggested that even among the small group who had chosen to buy Bitcoin, many people were not actually interested in having total control over their own money. Even the firmest advocates for Bitcoin's self-empowering potential, like Roger Ver, were entrusting coins to Mt. Gox and MyBitcoin, rather than holding the coins in their own addresses. And they were paying the price in lost and stolen coins. This raised questions about whether people really wanted, or were capable of taking advantage of, the decentralization that Bitcoin was offering. People may have trusted the code underlying Bitcoin, but they didn't necessarily trust themselves to deal with that code in the right way— and so they turned to outside experts to secure their money and make it easily available.

Meanwhile, the services that had become so popular in the Bitcoin community helped explain why governments and centralized authorities, like regulators, were often granted power in the real world. When people entrust money to financial institutions, they generally don't have the expertise or time to make sure the institution is doing its job. In most cases, it is much more efficient for people to band together and pool resources to ensure that their banks and exchanges are on the straight and narrow. Thus were created government agencies like the Federal Deposit Insurance Corporation, which backs up American bank accounts against losses, and checks to make sure that banks aren't putting deposits in danger.

Many libertarians and anarchists argued that the good in humans, or in the market, could do the job of regulators, ensuring that bad companies did not survive. But the Bitcoin experience suggested that the penalties meted out by the market are often imposed only after the bad deeds were done and do not serve as a deterrent. When it came down to it, in each case of big theft, Bitcoin users eventually went to government authorities to seek redress—the same authorities that Bitcoin had been designed, at least partly, to obviate. Mark Karpeles reported the Mt. Gox hack to the Japanese police and MyBitcoin users went to the FBI's cybercrime unit. Also not surprisingly, the police in these cases hinted that the Bitcoiners had created the mess and could clean it up themselves.

CHAPTER 11

●

November 2011

uccess was also testing the big ambitions and grand ideas with which Ross Ulbricht had started Silk Road.

After getting overwhelmed by new users in June 2011, he had brought the site back online, but on a more limited basis—with new registrations halted. His friend Richard, who had been helping him write the site's code, asked him: "Have you ever thought about doing something legitimate, something legal?"

Ross, in fact, had considered alternatives, and he began collaborating with Richard on a Bitcoin exchange—not a silly idea given the troubles that Mt. Gox was having. They began designing a prototype for their exchange while Ross continued running Silk Road.

In the fall, Ross was forced to consider his options seriously after a friend of his ex-girlfriend—the only other person who knew about his involvement with Silk Road—posted a revealing message on Ross's Facebook page: "I'm sure the authorities would be very interested in your drug-running site."

Ross immediately deleted the post and unfriended the woman

who had posted it. But he was terrified and went over to Richard's house to talk with the only other person who knew his secret.

"You've got to shut the site down," Richard told him, after Ross had arrived and explained what had happened. "This is all they need. Once they see this, they can get a warrant and it's over. This is not worth going to prison over."

Ross told Richard that he had, in fact, already sold the site to someone else, but Richard could tell Ross was still very shaken. And there was good reason for him to be: Ross had not sold the site. He lied to Richard as one part of his effort to cover his tracks. He was, in fact, still firmly in charge of Silk Road.

Looking at the numbers made it easy to see why Silk Road was a hard business to turn away from. In August alone, the site had generated $30,000 in commissions. There was so much business that in September Ross hired his first staff member to help him out—a user of the site who went by the name chronicpain.

More was involved than the money, though. Ross's site was actually accomplishing the big things he'd been dreaming about a year before—fulfilling both his ego and his ideals. On the Silk Road forums, he was able to give his grandiose aspirations free rein:

"We've drawn a line in the sand and are staring down our enemies. Like it or not, if you are participating here, you are standing on that line with us. This is not about making money. This is about winning a war. Look how far we've come in 8 short months. We are JUST getting started."

The notion that a site dedicated to selling heroin and forged passports was a moral cause would seem to many in the outside world an exceedingly bold claim. But for Ross, Silk Road was an application of the ideas advanced by the philosophers and economists whom Roger Ver and Erik Voorhees also loved—the ones who prized freedom above all else. According to this moral code, people should be allowed to do anything they please as long as it didn't hurt others. Freeing people from the constraints that held them back was an achievement of the highest order, even when all that it allowed was a junkie to get his fix.

The emphasis on freedom did not mean that Silk Road was an entirely lawless place. If a product, such as child porn, required the victimization of someone else, it was banned from the site—and immediately removed by Ross—following the one rule that all the anarchists and libertarians tended to agree on. When Ross created a category called forgeries, there were also limits: "Sellers may not list forgeries of any privately issued documents such as diplomas/ certifications, tickets or receipts," he wrote on the Silk Road forums. But documents created by governments were fair game.

The success of Silk Road was certainly offering Ross a freedom unlike anything he had experienced before. In late 2011, he sold his pickup truck and moved to Sydney, Australia, where his sister lived. All he needed for his job was his Samsung laptop. He would fit in his work around trips to Bondi beach, where he surfed and hung out with a crew of friends he quickly fell in with. As always, his cool gravelly voice and good looks made it easy for him to meet women. But he had, by now, learned his lesson about discussing Silk Road with anyone. When people asked what he did for a living, he would explain that he was working on a Bitcoin exchange. But for someone involved in such a bold and transgressive enterprise, Ross was a surprisingly fragile and sensitive soul. After a day of walking around Sydney with a girl he liked, just before the New Year, Ross explained how difficult his double life was becoming in the one forum where it was possible—the diary on his computer.

"Our conversation was somewhat deep," he wrote of his walk with the girl. "I felt compelled to reveal myself to her. It was terrible. I told her I have secrets. She already knows I work with Bitcoin which is also terrible. I'm so stupid. Everyone knows I am working on a Bitcoin exchange. I always thought honesty was the best policy, and now I don't know what to do. I should've just told everyone I am a freelance programmer or something, but I had to tell half truths. It felt wrong to lie completely so I tried to tell the truth without revealing the bad part, but now I am in a jam."

It was, though, the norm for Ross to fluctuate between self-doubt and hubris. The unusual combination seemed to actually be one of the keys to his success. His self-reflectiveness led him to question everything and constantly rework his site, while his confidence kept him going when he got down on himself.

Keeping his spirits up was becoming easier in late 2011 because Silk Road had attracted a lively community of users. Ross had also found someone he trusted on the site—a vendor on Silk Road who became a staff member and went by the name Variety Jones. Ross described him as "the biggest and strongest willed character I had met through the site thus far." Variety Jones, or vj, as Ross referred to him, pointed out flaws in the site's design and helped Ross figure out how to fix them. More important, he became a sort of confidant and even a friend to Ross, helping him think through the best way to run the site, and to feel less lonely.

When Ross was worrying about the people who might compromise him, Variety Jones came up with a clever idea: Ross could change his name on the site from silkroad to Dread Pirate Roberts. The name carried a swashbuckling panache that Ross liked, but it also provided something more important: an alibi. In the movie *The Princess Bride*, Dread Pirate Roberts was a name that was passed along between vagabonds. This could allow Ross to later say that the job of running Silk Road had been done by different people at different times.

"start the legend now," Variety Jones told him in a chat.

"I like the idea," Ross wrote back. "goes along with my captain analogy."

Variety Jones also helped Ross hone his public pronouncements on the site, which never showed any of the insecurity that Ross had in his real life. In his "State of the Road Address," posted on the Silk Road forum in January 2012, Ross explained that the site was "never meant to be private and exclusive. It is meant to grow into a force to be reckoned with that can challenge the powers that be and at last give people the option to choose freedom over tyranny."

If nothing else, Silk Road was indeed providing a good show-case for how anonymous markets and decentralized currencies could work in practice. In early 2012 Silk Road was still essen-tially the only place where people were regularly using Bitcoin to make real online, anonymous transactions—and the system was working as well as the Cypherpunks might have hoped. Silk Road customers were regularly sending payments of thousands of dol-lars—or hundreds of Bitcoins—to vendors on the other side of the world. In early 2012 there were vendors in at least eleven countries and many of them were willing to send their products across in-ternational borders. All of this was done using Bitcoin addresses and private keys that did not require either side to provide any per-sonal information. There were essentially no complaints on the site about the Bitcoin payment system, and many users who came for the drugs grew to admire the ways in which the virtual currency improved on existing payment systems. It turned out that when the incentives were high enough, lots of people, even those in altered states, could use Bitcoin as intended. The only occasional gripe was about the volatile price of Bitcoin, which made it hard to know how much a vendor would be charging a week later. But Ross dealt with this by creating a clever hedging program that allowed cus-tomers and vendors to lock in a price.

Silk Road was also providing a demonstration of how the market could work to keep an unpoliced community in check, even one where the members of the community went by screen names like nomad bloodbath, libertas, and drdeepwood. The primary

tool that brought accountability to this anonymous market was
the same sort of feedback mechanism used by eBay and Amazon.
When a customer received a Silk Road product through the mail,
he or she was asked to rate the transaction on a scale from 1 to 5.
Even if no one knew the real name of a seller, the reviews attached
to a seller's screen name would allow customers to determine if
that particular vendor was trustworthy. A few bad reviews could
sink a seller's business.

This feedback loop created a remarkably engaged online community in which pot and heroin highs were discussed with the
same level of analytical detail that *Consumer Reports* brought to
its toaster reviews. And it injected accountability into this apparently lawless land. An academic study of Silk Road later found that
nearly 99 percent of all reviews gave the maximum score of 5 out
of 5. This helped keep Silk Road growing, from 220 vendors in
late 2011 to around 350 in March 2012. The value of all sales in
the spring of 2012 was around $35,000 a day. Ross was taking be-
tween 2 and 10 percent of each purchase as a commission, depending on the size of the order. In March, that amounted to nearly
$90,000 in commissions, collected in Bitcoins.

There was, however, an often unspoken irony in the success
of Silk Road, and of Bitcoin for that matter. The site and the currency, which aimed to circumvent the power of the government,
were largely built on technology that had been created by the
government or through research sponsored by tax money. The
Internet itself was an outgrowth of several government research
programs, and the Tor network that served as a backbone of Silk
Road had been created by the Office of Naval Intelligence. Bitcoin, meanwhile, relied on advances in cryptography that had been
built thanks to government funding. Ross himself had gained the
expertise to build his government-eluding site after attending one
of the best-funded public high schools in Texas and two public

universities. It was no coincidence that these technologies did not emerge from a place with a weak government and bad educational systems. But Ross focused on the wrongs the government committed and ignored the advantages it had provided.

That same government was, of course, not going to sit by idly while the technology was used to support an online drug bazaar. Ross didn't know it, but in the fall of 2011 the Baltimore office of Homeland Security Investigations, or HSI, the law enforcement arm of the Department of Homeland Security, had opened accounts on Silk Road and began making small-scale purchases. This led federal agents, in January, to the doorstep of a young man in one of the poor suburbs of Baltimore who was known on Silk Road as DigitalInk. In real life, DigitalInk's name was Jacob George and he had been buying drugs—including methylone, bath salts, and heroin scramble—on the streets of Baltimore and reselling them online, becoming one of the most popular vendors on Silk Road after joining the site in July 2011.

After DigitalInk was arrested in early 2012, he immediately agreed to cooperate with the police. His record of Bitcoin transactions provided only limited information about the identity of his customers, owing to the lack of personal information connected to Bitcoin addresses. But it was a first strand of loose yarn for the officers to start pulling at. And in March the HSI bureau in Baltimore got approval from local prosecutors to form a task force, with other federal agencies, that would aim to burrow further into the cryptographically secured drug bazaar. The task force was given the name Marco Polo in deference to the man who explored the original Silk Road and all the new wonders it contained. A short while later, the agents in Baltimore created an undercover identity for themselves on Silk Road, with the screen name nob, and set out to build a relationship with a man they knew of only as Dread Pirate Roberts.

PART
TWO

CHAPTER 12

●

February 2012

After running away from the United States government to pursue his antigovernment vision, Roger Ver had chosen to live in a place that was uniquely unreceptive to his brand of antiauthoritarian politics. Japan was a country that was still deeply wedded to traditional hierarchies with an educational system that taught its citizens from a young age to obey authority. This was evident in the country's rigid business traditions—the bowing and exchanging of cards—and in the spiky-haired punks in Tokyo, who waited patiently for walk signals, even when there were no cars in sight.

Roger had picked Japan, not because it would allow him to be around other like-minded people, but because he liked the orderliness of Japanese culture—and the women. He had met his longtime Japanese girlfriend at a gathering in California and even she had almost no interest in politics. As Roger discovered, the deferential culture made Japanese people uniquely skeptical about a project like Bitcoin that aimed to challenge government currencies. Japan was the only place Roger had encountered where people's response, when he described Bitcoin, was to call it scary—rather than interesting or silly. This was due, Roger believed, to the way in which the virtual currency broke from the government's mandates about how money should work. One of the only people with whom Roger had gotten any traction in Japan was a local pornography tycoon.

Luckily for Bitcoin, Roger's job and wealth allowed him to wander far beyond Japan. In early 2011, he commenced his effort to renounce his United States citizenship so that he would not have to pay another dollar of taxes to support a government he considered immoral. Japan, with its sense of tradition and history, made it almost impossible for foreigners to gain citizenship, so Roger made plans to travel to Guatemala to start the process of applying for citizenship there. He was also traveling constantly for his work with Memory Dealers—looking for cheap hardware—and everywhere he went he would talk about his new passion. While visiting the Chinese manufacturing hub of Shenzhen, he held the first-ever Bitcoin Meetup in China and paid for the group meal himself. Whenever he ended up in a taxi, he would set up his driver with a smartphone wallet and try to pay his fare in Bitcoin. When Roger began looking for an engagement ring, he promised the online diamond merchant BlueNile that he would buy a $50,000 diamond if the company began publicly accepting Bitcoin (BlueNile ultimately demurred). He continued using his own company, Memory Dealers, to promote Bitcoin by offering discounts to people who paid with Bitcoin, and by selling the popular "physical Bitcoins," known as Casascius coins, manufactured by a man in Utah. Bitcoins, of course, have no physical quality—they are nothing more than an entry on a digital ledger. But the creator of the Casascius coins printed the private key for an unspent Bitcoin on the inside of a hologram, attached to a specially manufactured coin with the Bitcoin emblem. A person could spend the Bitcoin by peeling off the hologram and using the private key. These Casascius coins would later become the most widely used image of Bitcoins when news organizations needed a picture of something to accompany stories about the virtual currency.

When Roger got into conversations about Bitcoin, he had a few stock lines he would deliver, always with the same crisp elocution and conviction—almost as if he were in a reverie.

"I'm pretty confident that Bitcoin is *the* most important invention since the Internet itself. The world is changing because of Bitcoin right in front of our eyes and it's such an exciting time to be a part of this," he liked to say. "I've been spending just about every waking moment focusing on Bitcoin."

Roger had always been a good salesman in part because of his ability to communicate his own conviction, but also because he had an intuitive sense for what people wanted and knew how to meet them at their level, without demanding agreement with his beliefs. His pitch for Bitcoin to the antigovernment activists emphasized the ability to buy drugs with Bitcoin, even though Roger himself was an abstainer who had never smoked a cigarette. When other Bitcoiners said that Roger's talk of drugs and dodging taxes could tarnish Bitcoin's reputation, he replied that he always adjusted his arguments to his audience.

"If I was going on the Oprah Winfrey show, I should certainly use a different list of talking points," he explained on the Bitcoin forum.

Roger, then, had the rare resources and abilities to help sell Bitcoin beyond the small fringe communities where it had so far been cloistered. And he was dedicating his life to doing just that. In addition to the personal pitches and purchases, he was eagerly supporting any companies he could find that might help expand Bitcoin's appeal beyond libertarians and heroin addicts. He gave $100,000 to Jesse Powell, his old friend who had come to Tokyo to help out with Mt. Gox. Jesse had been so struck by Mark

Karpeles's weaknesses that he decided to start his own exchange. But Roger's most significant investment early on would prove to be the one he made in a young New Yorker named Charlie Shrem. Roger had first seen Charlie talking about his company, BitInstant, on Bruce Wagner's *The Bitcoin Show*. A small, cherubic twenty-two-year-old, with a Brillo Pad of curly hair and a slight Brooklyn accent, Charlie pitched BitInstant as the easy way to get money into and out of Bitcoin without wiring funds internationally to Mt. Gox's bank account in Japan.

Roger quickly reached out to Charlie by Skype, and asked how much money he needed. Charlie offered him 10 percent of the company for $100,000. Roger sent over a wire payment for $120,000.

THE YOUNG MAN Roger had invested in was, outwardly, an unlikely candidate to become the entrepreneurial leader in a futuristic global movement like Bitcoin. He had grown up in the Midwood section of Brooklyn, in a Syrian Jewish community where all the kids went to the same religious schools. From early on, Charlie had struggled with social acceptance. He had been born cross-eyed and, after surgery to fix the problem, had to wear thick glasses. He was almost always the shortest one in his classes. As with so many other techies, Charlie's real-world struggles led him to cultivate an active life online, where he knew many of his friends by their screen names.

But a surprising confidence lurked beneath Charlie's anxious exterior. As the oldest child and only son in a family with four sisters, he was treated like a prince by his mother. He had discovered that while other kids could be difficult to win over, grown-ups were generally an easier audience. He was the one kid at his synagogue who would go up and shake the rabbi's hand after services and his energy and good spirit generally appealed to adults. As he grew up, he found his personality lent itself naturally to business, which was highly valued in his community and in his family; his parents ran their own jewelry businesses. When he was a freshman at Brooklyn College, he and a few friends had founded an online deals site, somewhat like Groupon. He blossomed into a confident salesman when pitching his ideas.

Charlie had initially learned about Bitcoin through an article about Silk Road. He had gone on the forums and found another user who was thinking about launching a deceptively simple startup: a company that would make it easier to get dollars into and out of Mt. Gox. The man, Gareth Nelson, lived in Wales and had already programmed a prototype. Charlie confidently pitched what he could bring to the project, telling Gareth that he knew people at PayPal—"very high-up"—and would call to get their support. In reality, though, the first people Charlie got help from were his parents. Still living in the basement of his childhood home in Brooklyn, Charlie asked his mother if she would be willing to give him a seed investment. Charlie's mom, who ran the jewelry company Bangles by Kelly, rarely said no to her only son and didn't disappoint him this time, transferring $10,000 to him.

Charlie was a departure from the idealists who had been driving Bitcoin development so far. His first-ever post on the Bitcoin forum was not about the power of decentralization but an offer to sell JetBlue airline vouchers for Bitcoins. Over the next months he would offer magazine subscriptions, "Fuzzy Toe Socks," and throwing knives.

It turned out that Charlie's willingness to throw things at the wall, to see if they would stick, was not a bad thing at this point. The idealists who had been driving the Bitcoin world often got caught up in what they wanted the world to look like, rather than figuring out how to provide the world with something it would want. The business model being pursued by Charlie and Gareth was designed with the very practical aim of making it easier for customers to get Bitcoins than it was to get them from Mt. Gox, which required wiring money overseas and placing orders on the exchange. Just as Charles Schwab dealt with the New York Stock Exchange so that its customers didn't have to do so, BitInstant handled all the dealings with Mt. Gox, making the process of acquiring Bitcoins faster and easier.

Charlie's swagger led him to generate ideas, and act on them, in a way that was still unusual in this young industry. But his confidence also came with a recklessness that would become a liability. On the Bitcoin forum, Charlie advertised his love of marijuana and offered Silk Road users help and advice. Less publicly, he began working with a Florida man who helped Silk Road users get Bitcoins to buy drugs. Charlie was smart enough to include a section on the BitInstant site about the company's intolerance for anybody using Bitcoin illegally and he chose not to advertise his own company on Silk Road. But when a Florida man, who went by the screen name BTC King, approached Charlie about privately exchanging large amounts of money for Silk Road customers, Charlie devised a way to do it without attracting notice. When Charlie's programming partner in Wales questioned Charlie about the deals with the man, Charlie argued that they wouldn't be a problem.

"He has not broken any rules and silk road itself is not illegal," Charlie wrote to Gareth. Besides, he said: "We make good profit from him."

WHEN ROGER VER invested in BitInstant, he could tell that Charlie was a raw talent and offered himself as the company's marketing director to help steer Charlie's idea. He then connected Charlie with Erik Voorhees. Erik, who was still living in New Hampshire, was more ideological than Charlie, but he was also more careful and grounded, and Roger thought they would complement each other. The month Erik joined BitInstant, the company processed $530,000 in transactions, up from $250,000 just two months earlier.

As they began working together, Roger and Erik jokingly gave Charlie the nickname "Statist" for his more traditional politics and respect for government. But that didn't stop BitInstant from becoming a popular service among all the ideologically motivated people whom Roger and Erik were winning over, who were looking for the easiest way to get their hands on Bitcoin.

In February Erik appeared at Liberty Forum—one of the Free State Project's two major annual events—to speak about Bitcoin's appeal to anyone opposed to the American government. The room was packed and Erik was mobbed afterward by interested people wanting to get involved. The price reflected that interest. After bottoming out in late November at around $2, by February the price of a single Bitcoin was stabilizing around $5. It didn't hurt that Bitcoin made its first serious foray into popular culture in January 2012 when an entire episode of *The Good Wife* was based on a plot about Bitcoin.

In April Erik traveled from New Hampshire down to New York to meet Charlie in person for the first time and to make a presentation at the first-ever New York Tech Day, an event designed to connect startups and investors. Charlie and Erik spent the morning setting up their booth at the storied Park Avenue Armory with slick BitInstant banners and branded key chains.

Soon after the doors opened, two older gentlemen with the casual whiff of money approached Charlie. He launched into his elevator pitch for Bitcoin, leaving out anything about central banks, and focusing on the ability to transfer money around the world free. The two men had never heard of Bitcoin, but one had worked in the import-export business and knew how expensive it could be to move money across international borders. What's more, they liked Charlie's irrepressible energy, which was immediately evident, and recognized from his last name, Shrem, that he was a member of the tight-knit Syrian Jewish community that they belonged to.

On the spot, the two men offered Charlie a free space to work at The Yard, an office for startups they had recently opened in Brooklyn. They also suggested they would be interested in making an investment in BitInstant. That same afternoon, Charlie visited The Yard, built out of an old industrial building in the hip neighborhood of Williamsburg. Bitcoin was quite literally moving out of the basement and into the real world.

WHEN CHARLIE HAD begun BitInstant less than a year earlier, it was a response to a very specific and narrow problem—the difficulty of getting money into Mt. Gox's bank accounts to buy Bitcoins. But Charlie's conversation with the two potential investors at New York Tech Day illustrated his growing awareness that his company could also help ordinary people take advantage of a much more practical service than Bitcoin could offer the world. Thanks to his upbringing in a community of entrepreneurs, Charlie knew that in 2012 businesses still had few good ways of instantly transferring money to pay for goods and services. A normal bank payment took several days, and a wire transfer moved faster but cost $30 to $50 each time.

Charlie's practical bent had led him, unwittingly, to an issue that had rarely been a part of the Cypherpunk discussions but that was perhaps the most widely acknowledged problem with the existing financial system: the creakiness of the old payments system.

In March 2012, a month before Charlie found his investors, the Federal Reserve had held a daylong conference about consumer-payment systems at which there was a lot of grousing about the fact that despite all the technological innovation going on in the world, the infrastructure for moving money around the country was still based on technology from the 1960s and 1970s. The Automated Clearing House, or ACH, which facilitated payments between bank accounts, was created in the 1970s and had not changed much since; this helped explain why bank transfers took at least a day to go through. For most Americans, the easiest and fastest way to send money to a friend or family member was still the old-fashioned paper check. This problem was not just in the United States. A week before New York Tech Day, the Canadian government announced the launch of a new digital currency effort, called Mint Chip, that it hoped would spur innovation in payments.

The weakness of the existing system had been evident during the financial crisis when the Wall Street bank Morgan Stanley needed a $9 billion infusion from a Japanese firm. The agreement was reached on a Sunday, but the money could not be sent because the wire network was down for the weekend and the next day was Columbus Day. It turned out that even banks couldn't send each other money on holidays. In order to get around this, the Japanese bank cut an absurd $9 billion paper check.

With Bitcoin, transfers did not happen instantly, as was sometimes claimed. A Bitcoin transaction was official only after it had been confirmed by a miner and included on the blockchain, which generally took a minimum of ten minutes. But it took around ten minutes at any hour on any day of the week and could be done from a smartphone, which was a lot better than waiting until Tuesday.

The potential of the Bitcoin network as a new, cheaper, and faster payment system represented an opportunity for the network that went beyond the controversial anonymity it appeared to offer, and the ideological attraction of its decentralization. Charlie wasn't the only person who had spotted this opportunity. Two former fraternity brothers at Georgia Tech had founded a company called BitPay, which looked to harness the network as a cheaper way for merchants to accept online payments, while also giving Bitcoiners a place to actually spend their virtual currency. With BitPay, merchants could accept Bitcoin, and BitPay would immediately convert the virtual currency into dollars and deliver those dollars into the merchant's bank account. This was attractive to merchants because BitPay charged around 1 percent for its service while credit card networks generally charged between 2 and 3 percent per transaction. What's more, whereas credit card companies could recall money from a merchant in the case of a customer dispute, Bitcoin transactions were irreversible.

The opportunity here was also evident to another businessman from Charlie's Syrian Jewish community, a man named David Azar, who was the son of Charlie's childhood dentist. When David heard about Charlie's business from a friend, he was intrigued. David ran a chain of check-cashing shops and he had intimate experience with all the drawbacks of the existing payment networks.

David, an energetic entrepreneur who came across to others as something of a street fighter, invited Charlie to his office, which was just a few blocks from the BitInstant offices. In their first meeting, David boldly told Charlie that he wanted to invest money in Charlie's company and had the money to do it. Charlie was thrilled, but explained that he was already working with two other investors from the Syrian Jewish community who were planning to put money into BitInstant. David made it clear to Charlie that he wanted to make the investment on his own and that he was not one to easily take no for an answer.

CHAPTER 13

●

May 2012

Less than a year earlier, when Charlie Shrem had stopped by
the first Bitcoin conference in New York, he had been too
timid to introduce himself to anyone. Now, in the early
summer of 2012, he was the toast of the Bitcoin world and was
getting invitations from all directions. In late April he flew to San
Francisco to appear on a panel about the future of money. In the
crowd afterward, a small, svelte Russian man introduced himself
and asked if Charlie would be interested in traveling to Vienna to
join a small group advising the man on his own Bitcoin startup, a
credit-card-thin device that could serve as a Bitcoin wallet. Once
Charlie was back in New York, he discovered that the man, Alexander Kuzmin, was a minor Russian tycoon who was directing a
fortune he'd made from Siberian oil to anarchist causes. Kuzmin
also invited Erik Voorhees, Roger Ver, and Gavin Andresen to come
to Vienna and sent along Bitcoins to pay for their travel expenses.

While Charlie and Erik prepared for the trip, they were also
being pursued by the two investors who wanted to give $1 million
to BitInstant. This was surprisingly hard for Charlie because of

his instinctual aversion to telling people things they didn't want to hear. Instead, he strung them both along. When the first investors had to cancel their plans to join Charlie in Vienna, the second, David Azar, quickly booked a ticket. In Vienna, when the Russian mogul wasn't pampering the Bitcoiners at his airy two-story penthouse, David treated the BitInstant team to a good time. The men visited a sex club that had a hefty entry charge and an additional fee for each intimate act with the women. After paying the admission fee for the others, David turned around and went back to his room. David also quietly offered Charlie several thousand dollars on the side if Charlie chose David's investment.

When Charlie and Erik returned to New York they decided to go with David. This required a surreptitious exit from the working space that had been given to them by the first potential investors. While Charlie broke the bad news, Erik hurriedly moved all their computers into Charlie's BMW so they would be ready to leave in a hurry when Charlie left his meeting with the disappointed men who had put their hopes in him. Charlie got yelled at but, as he and Erik sped away laughing, it felt like just another exhilarating incident in their intoxicating ascent.

Eric was becoming a figure in the Bitcoin world in his own right, thanks in no small part to a gambling site, SatoshiDice, which he had started up in late April. The game of odds was based on the same hash functions and math underlying Bitcoin, and the outcome of each bet was visible on the blockchain. Players gambled by sending small payments to specific Bitcoin addresses, and winning bets immediately paid out. If this had been done using traditional payment networks, the transaction fees would have made it prohibitively expensive, but with Bitcoin the payments could go in and out free. The game itself had been invented by someone else, but Erik bought the concept for 45 Bitcoins, gave it a user-friendly website, and got it up and running. By July it had already become wildly popular and he began making plans to sell stock in the company on a Bitcoin stock exchange set up by a man in Romania.

Erik made his commitment to Charlie and BitInstant more firm when he moved to New York full-time in July and convinced a friend from Colorado, Ira Miller, who had been working with him on Bitcoin projects, to move with him. The BitInstant crew worked briefly out of Erik and Ira's new apartment in Brooklyn, but they soon rented their own office in Manhattan just feet from the storied Flatiron Building. Charlie had his own office with windows looking onto the street. In the main room, he installed a big screen that displayed the live price of Bitcoin.

To Erik's delight, Charlie was beginning to be won over by the more ideological arguments for Bitcoin. During the summer, they met Roger in New Hampshire for PorcFest, a festival in the woods held by the Free State Project. They were amazed to find that many of the vendors already accepted Bitcoin, allowing them to make it through the weekend using almost no dollars. They had made the theme song for BitInstant—"It's Yo' Money Why Wait?"—and Erik would occasionally blast it from the back of his Subaru Impreza.

IT WAS, THOUGH, becoming increasingly clear that Bitcoin was on a trajectory that was going to be hard to sustain as the authorities became more aware of it.

Silk Road was still driving a significant portion of the real transactions on the Bitcoin network, including many of the people buying coins from BitInstant and Mt. Gox. When a friend asked Charlie about Silk Road, Charlie explained that "it funds a decent percentage of the overall Bitcoin economy."

The consequences of this had become hard to avoid when a remarkably well-informed report entitled "Bitcoin Virtual Currency," prepared by the FBI, had leaked in May. From the first line, it was evident that the FBI did not generally view Bitcoin in a positive light; the report described the network as a "venue for individuals to generate, transfer, launder, and steal illicit funds with some anonymity." The report also said that the agency "assesses with medium confidence that law enforcement can identify, or discover more information about malicious actors."

Charlie kept working with the BTC King, who helped Silk Road customers acquire coins. But Charlie was increasingly trying to follow the relevant rules when it came to gathering information about customers who made transfers above prescribed minimum amounts. He also registered with the Treasury Department agency responsible for regulating money transmitters, the Financial Crimes Enforcement Network, or FinCen.

The issue of Bitcoin's reputation had been a steady topic of conversation when Charlie, Gavin, and the others had been in Vienna. At a café Charlie had chatted with Gavin about some sort of foundation that could serve as a neutral voice to bring the technology into the mainstream and create some distance from Silk Road.

When Gavin returned from Vienna he had connected Charlie with Peter Vessenes, a Seattle entrepreneur who was trying to break into the Bitcoin space. Peter did not have much of a business plan, but he had some practical business experience and had already managed to land some funding for his startup, CoinLab. He was also very eager to help Bitcoin break into the mainstream.

In a series of increasingly excited e-mails, Charlie and Peter both emphasized the need for a foundation that could separate itself from the virtual currency's controversial past. Charlie told Peter that those involved had to be people "without tarnishes and have spotless reputations within Bitcoin. Anyone involved with even an inkling of mistrust ruins our whole legitimacy."

Roger was included in the planning of the foundation, and promised to donate 5,000 Bitcoins to support it. But it was decided

early on that Roger would not take a seat on the board because of
the prison term he had served. Peter pushed to be given a leadership
role because of his past entrepreneurial experience. When Charlie
and Roger suggested that others—such as Jed McCaleb and Jesse
Powell—be included, Peter quickly shut down the idea, saying it
would be better to restrict the planning to a small circle of people.

The man who would serve as the glue in bringing this all together was Patrick Murck, an unassuming Seattle lawyer whom
both Charlie and Peter had independently found. Patrick had not
come to Bitcoin with the same intentionality as so many members
of the early community. He had spent his first years out of law
school working at a firm in Washington, DC, where he had grown
up as the child of a federal employee.

Recently, though, not long after his son was born, Patrick's
mother-in-law was diagnosed with cancer, and he and his wife
had sold everything and moved to Seattle to help care for her. His
wife had given up her own job at the National Wildlife Federa-
tion. Patrick had begun to get his professional life back on track in
Seattle by getting a job at an advertising startup that focused on
digital games and tokens; there he began to learn about the law
surrounding digital money. When that job didn't work out, Patrick
found that he was one of the only people with any legal expertise
in anything close to virtual currencies and he started consulting
for Bitcoin startups like BitInstant.

Patrick was indicative of the increasingly practical turn that
Bitcoin was taking. He was not a libertarian—he had, in fact, volunteered for the Obama campaign in 2008. His work with Bitcoin
had started as a job and evolved into a passion, rather than the
other way around.

In a first group meeting, by phone, the men all agreed that the
foundation would steer clear of the politics that had been associated
with the technology and would, instead, focus on standardizing

the technology and providing a neutral meeting ground for the community. They held out, as their model, the foundation connected to the open source Linux operating system. Occupy Wall Street this was not.

All the men on the call were aware that one of the biggest complications that faced them was Mt. Gox. The exchange had continued to be the largest venue for buying and selling Bitcoins. Mark Karpeles had brought on new staff, many of them fellow French expatriates, and found the company larger offices just a few blocks from Roger Ver's apartment (so close, in fact, that Mark's staff initially used Roger's wifi network). But Mark's social skills had not grown with his company. Despite having a Japanese wife and now a young son, he rarely talked about them with others and seemed much more interested in his cat Tibanne, about whom he posted loving items on Twitter and YouTube. At work, Mark kept all the important responsibilities in his own hands and as a result the business moved only as quickly as Mark did. The exchange was constantly facing complaints about long wait times and poor management. When Roger lent Mark money, he had trouble getting paid back, and when he needed a transaction to go through, he would sometimes have to visit the Mt. Gox offices.

Peter Vessenes, in Seattle, was hoping to raise money from investors to either purchase Mt. Gox or take over some of its management. Peter had written to Mark and told him: "My gut, and it's just a gut feeling, is that Gox could use more finance and global business experience to grow in the way you guys want it to."

At the same time Peter was planning a first in-person meeting for the group behind the Bitcoin Foundation, he also made a trip to Tokyo to sell Mark on the idea of teaming up. Personally, the men were like oil and water. Peter was the genial American businessman who liked to ease into business conversations by talking about family and personal life, while Mark rarely discussed anything beyond work, and hardly even that. By the end of the visit, though, the men had begun planning for Peter's company to take over Mt. Gox's American customers. Peter did not invite Mark to a first meeting of the group behind the foundation, but he did secure a promise from Mark to donate 5,000 Bitcoins to the organization. He also got Mark to hand over the domain name BitcoinFoundation.org, which Mark had acquired a year earlier.

Almost as soon as Peter was back from Tokyo, Roger, Gavin, and Charlie flew to Seattle for a meeting to formalize the Bitcoin Foundation. During the two days of meetings, Gavin made it clear that he had no interest in doing anything other than working on the Bitcoin code. Charlie, meanwhile, was eager to take charge of the foundation's annual conference, which he said could raise $200,000 or more. Patrick Murck, the lawyer, took on much of the hard work of bringing the foundation into existence.

To underscore the decentralized principle of Bitcoin, the group agreed that the bylaws for the foundation would be posted on GitHub, the open source software site, where people could comment and suggest additions or changes. But in a rather undemocratic step, the men in Seattle decided to anoint themselves, and Mark Karpeles, the initial members of the Bitcoin Foundation board. Peter had the clever idea of including, as a founding member, Satoshi Nakamoto, or whoever could prove ownership of Satoshi's public key: DE4E FCA3 E1AB 9E41 CE96 CECB 18C0 9E86 5EC9 48A1.

A rare tense moment during the gathering came when Roger dressed Charlie down for constantly opening up his laptop to deal with small tasks at BitInstant—transferring money or dealing with customer e-mails. For Roger, this brought back bad memories of Mark's inability to delegate responsibility to others. Roger, with evident frustration, told Charlie to hire more people to take care of things for him.

"You are the CEO," Roger said. "You shouldn't be responding to customer service requests."

The eager-to-please Charlie put his laptop away, but he had trouble keeping it closed.

At the end of the day, the group retreated to the palatial waterfront home of another cofounder of Peter's new virtual currency company—a former Microsoft executive—who lived on a beautiful, exclusive peninsula near Seattle. When the wealthy neighbors wandered over, Roger immediately got them all set up with Bitcoin wallets on their phones. Watching Roger evangelize with his usual gusto about "the most important invention in history since the Internet," Charlie said to the others, with a laugh: "Look at Bitcoin Jesus." It was a nickname that would stick.

The luxurious evening on the water made it clear that Bitcoin was losing some of its fringe appeal but winning some useful friends.

CHAPTER 14

●

August 2012

Charlie Shrem and Erik Voorhees walked along the southern edge of Madison Square Park to Benvenuto Café. They were there to meet Barry Silbert, one of the big names in the New York tech scene. As Charlie walked into the café, he expected to see the sort of brash icon of new money caricatured in movies like *The Social Network*. What he found instead was someone with a boyish face and straight bangs that made him look almost as young as Charlie.

Only thirty-three, Barry Silbert was viewed as a prodigy in the financial industry, having worked at a Wall Street firm, managing bankrupt companies, before leaving to create a financial startup that had made it easier to buy and sell the stock of companies that didn't trade on stock exchanges. The company, SecondMarket, landed Barry on every list of forty under the age of forty.

Barry had already been quietly exploring Bitcoin for months. His interest was not political. He saw Bitcoin's potential to address inefficiencies in the existing ways of moving payments and other elements of the existing banking system. Given the fringe status of Bitcoin, Barry feared that going public with his interest in the technology could damage the reputation of his company, which was funded by several leading venture capitalists. But behind the scenes, Barry was seeking out anyone who could connect him with interesting virtual-currency investments. He had also spent around $150,000 buying up Bitcoins over the course of 2012.

Charlie and Erik were eager for the meeting because David Azar had proved difficult to pin down since the BitInstant guys agreed to accept his investment after the trip to Vienna earlier in the summer. If nothing else, Barry could advise them on how to handle the situation.

Barry obliged, but he also saw an opportunity for himself. He had already made a few angel investments in Bitcoin companies with his own money—including one in CoinLab, the Seattle company founded by Peter Vessenes—and was eager to expand his portfolio. What's more, he had recently gotten one of the biggest venture capital firms in New York—one of his early investors—excited about Bitcoin.

Within days, Charlie had scheduled a coffee with Barry's big investor, Larry Lenihan, a partner at the billion-dollar firm First-Mark Capital, which had bet on startup stars like Pinterest and Aereo. When they met, Larry was slightly put off by Charlie's untamed energy and hubris, but he liked the idea behind BitInstant enough that he immediately contacted Barry and said he wanted to explore making an investment. In an e-mail to Charlie, he asked when Charlie and Erik could come in to meet his partners: "I'd also like to bounce an idea off you guys about having NYC invest—this could be important. It would be out of Mayor Bloomberg's office and it would provide enormous amounts of credibility for the effort."

•

DAVID AZAR'S OPPORTUNITY to invest in BitInstant was about
to disappear when he went with some friends to the Spanish island
of Ibiza. While lounging at Blue Marlin, one of the trendy island's
most famous beach clubs, David noticed two tall men with waves
of glossy brown hair, who would have drawn his attention even if
they weren't Tyler and Cameron Winklevoss.

The Winklevoss twins had become a cultural phenomenon owing
to their involvement with Mark Zuckerberg when they were all un-
dergraduates at Harvard. Zuckerberg had initially teamed up with
the brothers to build a social networking site, but when Zuckerberg
went off on his own and created Facebook, the twins sued him, claiming he stole their idea. They eventually won a $65 million settlement
and the story inspired the Oscar-winning film *The Social Network*.

Aware that the brothers were tech savvy and wealthy, David
seized the opportunity. He sidled up to Cameron and dropped the
name of a friend of theirs in New York. David then asked Cameron if he knew anything about virtual currencies. Cameron did
not and David's brief description elicited an interested nod. The
encounter ended with David promising to follow up.

David caught the brothers at an opportune moment. Recently
retired from their rowing careers, which had taken them to the 2008
Beijing summer Olympics, they were using their money from the
Facebook settlement to set up their own technology investing firm.
Just before they had left for Ibiza, Winklevoss Capital had leased an
office a few blocks from the BitInstant offices in Manhattan.

At their family beach house on Long Island the next weekend,
Cameron read over the articles David sent along. Both brothers
had majored in economics at Harvard and, after just a bit of reading on his laptop, Cameron called his brother over.

"You've got to come over here and check this out," Cameron said to Tyler.

Tyler always played the right-handed, rational check to his more dreamy, left-handed brother. But as Tyler began reading, he saw what his brother was talking about. Both realized this was either a scam or a big deal—but worth exploring. When they got David on the phone, he told the twins about the company he was preparing to invest in and offered to connect them with the guys at BitInstant, with the clear implication that the brothers might want to invest in it as well.

Two days later, Cameron arrived at BitInstant's headquarters on Twenty-third Street and folded his big frame into Charlie's office. The conversation with Charlie and Erik about how the blockchain worked and how Bitcoin was different from previous digital currencies that had not taken off—like Facebook credits—lasted for almost two hours. Charlie came across as something of a Tasmanian devil, with energy shooting in all directions, not always in an ordered fashion. But for every skeptical question the twins asked, Erik had a well-thought-out answer. Cameron was particularly impressed by Erik's decision to take his entire salary from BitInstant in Bitcoin and to keep his savings in the virtual currency. Within a few days, the twins let David know that they were prepared to invest alongside him and set up a dinner to work out the terms. With Charlie and Erik, they opened up a jokey banter by e-mail as the twins went back and forth about the basics of Bitcoin and the nature of money.

Cameron: "Money does have some intrinsic value, for example if you were freezing on top of a mountain and all you had was cash you could burn it to keep warm a la Cliffhanger."

Charlie: "Anything is valued differently in different circumstance. . . . A dollar bill to a coke head is worth more than a dollar bill to you and I."

Cameron: "What about a dollar bill to a stripper?"

Charlie and Erik were now back in the enviable but awkward position of being courted by two different investing groups.

Each member of the BitInstant team weighed in. Roger was not excited about the Winklevoss twins. He thought that they were free riders, who had gotten rich thanks to the legal system, rather than by inventing something real. He also worried about the terms of the deal that David and the twins were offering, which provided much less flexibility and gave David more control than most startup investors have.

Roger was still a libertarian, but he was a practical one, and he understood the value of money from established venture capitalists like Barry Silbert and FirstMark Capital and especially the value of getting some buy-in from the City of New York.

"This is one of the most interesting investors possible because I suspect it would give us a great deal of added protection against future trouble with regulators / financial police," Roger wrote from Tokyo, advocating for Barry and FirstMark over David Azar and the twins.

Barry was already taking Charlie and Erik under his wing and trying to soften some of their rough edges. He cautioned Charlie to stop making comments in his e-mails about his drinking and carousing. After taking the BitInstant guys to an industry party he wrote a laundry list of their social faux pas that they needed to work on:

Take it easy with name dropping . . . Larry would not appreciate it if he knew you were telling people he was buying Bitcoins.

Charlie—your defense of Bitcoin to Brian at Tribeca came across as very aggressive. Be patient, LISTEN and try to disarm each one of their arguments.

Do your best to keep your phones in your pocket. It is
anti-social—borderline rude—to be doing emails, twitter,
etc. during dinner.

Charlie didn't love the paternalistic guidance. But more important, when Charlie considered which investment to take,
David had something that Barry could never match: he was part
of Charlie's tight-knit Syrian Jewish community. On hearing that
BitInstant was thinking about taking an investment from Barry
Silbert, David exploded, accusing Charlie of disloyalty. Members
of the Syrian Jewish community generally viewed themselves as
having more responsibility to each other than to the outside world.
This was an insular community in which even marrying a Jew
from Europe or Turkey was considered intermarriage. Charlie was
terrified that he would become a persona non grata in his neighborhood if he backed out of his deal.

In addition, David's partners, the Winklevoss twins, had a
glamour that was hard for him to resist. To someone who had
always been the last one picked for dodgeball, the tall blond
Olympians promised not just money, but a life in which he could
no longer be ignored.

Then there was the danger of turning down David's money for
a deal with FirstMark that was only in the initial stages. Charlie
wrote to Barry:

Is it worth risking a good deal I have now to see if a deal
may or may not happen? I mean, everything up until now
with Larry has just been talk. I've been farther with other
VC's who flaked on me last minute. This deal I have now
has been in the works since June, 4 months and Im tired!!

Barry pushed back hard:

This is your company and you gotta do what you gotta do, but just want to throw in my two cents. It would be game changing for your business and the Bitcoin industry for FirstMark capital to make an investment in BitInstant.

From Tokyo, Roger struck up a back-channel conversation with Barry, both to explain what was holding Charlie back, and to see if Barry could make an offer that would put some of Charlie's concerns to rest:

Charlie currently feels some cultural pressure to close the other deal, but if your offer is better, he will have every reason to not accept it and won't have any ramifications from his social circle.

Barry agreed to put up a $75,000 convertible note in order to create a bit of breathing room while he and FirstMark worked on a more formal offer. Roger quickly wrote to Charlie: "I don't want to burn any bridges, but I don't think we should feel bad asking David to wait an extra two weeks. He has already demonstrated that he is not in a hurry by taking months and months to put together a deal."

Charlie did hold off, but he eventually resolved the issue between David and the twins on one side and Barry and FirstMark on the other by getting David to soften up some investment terms that had turned Roger off. Charlie also convinced Erik that David's experience in the check-cashing business would immediately help BitInstant deal with regulatory issues it could face as lawmakers looked to rein in virtual currencies. To close the deal with David, Charlie offered Erik a 2 percent stake in BitInstant. They finalized everything sitting on the porch of David's lawyer in the Syrian Jewish section of Brooklyn. The agreement gave Maguire

Ventures, the investment entity created by David and the Winklevoss twins, 22 percent of BitInstant for $880,000. Charlie kept 29 percent of the company and Roger kept 15 percent, with the rest being split among the other employees.

By the time the final contract was signed, Charlie was already reaping the most immediate benefit of the deal: he was serving as a personal Bitcoin guide for the Winklevoss twins. He began buying them coins and helped them use Bitcoin to pay a Ukrainian programmer for work on the Winklevoss Capital website. Charlie and Erik also set up a time to sit down with the twins and give them a more in-depth Bitcoin tutorial at their offices.

Charlie and Erik deliberately scheduled the meeting on a Saturday evening, when the conversation might bleed into a night of partying with the brothers, and the twins didn't disappoint them. After a session on Bitcoin, leavened with alcohol, Cameron invited Charlie and Erik to join him for a night out. Girls the twins knew showed up and the crew headed to a party in a loft downtown, followed by a visit to a speakeasy. Charlie got so drunk on shots of rum that he threw up on his shoes in the middle of the bar. He still managed to end up back at Cameron's apartment with a girl—though Charlie ruefully reported that it didn't go anywhere.

"What a night," Cameron wrote to Charlie and Erik the next morning. "I trust u guys made it home in one piece."

"That was a blast," Erik wrote back. "I had to peace out before I drowned in liquor."

It wasn't just Charlie and Erik who found all of this thrilling. For the twins, despite their past successes, investing in Bitcoin at this point still felt like getting in on the ground floor of something huge, before anybody else had even heard about it.

But before they had a chance to savor it, the first signs of trouble appeared. In early November, a hacker attacked the BitInstant site, forcing it down several times. The hacker demanded a 10,000

Bitcoin ransom, keeping Charlie's small team of programmers working around the clock. At about the same time, BitInstant's longtime bank, Citi, began asking hard questions after months of not paying much attention to the account. This forced BitInstant to temporarily stop taking in new money through its bank account. A little more than a week after the investment was made official, David Azar snapped at Charlie: "I didn't sign up for this."

As David took ownership of the company, he questioned why the business wasn't growing faster. At the same time, he declined to hand over the first tranche of money. He demanded a full audit of BitInstant that was taking much longer than expected. He would shoot off brief e-mails like machine-gun fire, asking new questions before he got answers to the previous ones.

Erik watched, with a mixture of fascination and fear, as the arguments between Charlie and David quickly took on the form of a feud between angry siblings.

"David, I don't appreciate you calling me a child and speak to me in a condescending manner. I've always treated you with the utmost respect and I would think I deserve the same from you," Charlie wrote in an e-mail to David in early November after one fight. He went on:

> To this date, you still have an elementary level of Bitcoin and BitInstant. I need you to understand *why we are in business*, and what we are trying to accomplish in this world. You tell me that I need to learn everything from you, well *you still have not learned anything from us*.
>
> You need to make a decision on how you want to act going forward, with your attitude and position towards us.

CHAPTER 15

October 2012

In mid-October Charlie Shrem and Erik Voorhees played host to Wences Casares, a slender man with dark movie star looks, a sophisticated accent, and clothes that signaled elegance and ease. Wences had reached out to the BitInstant team out of the blue, giving little indication of his specific intentions regarding Bitcoin. As he began talking with Charlie and Erik, though, he quickly came across as very different from the previous curious programmers and entrepreneurs who had stopped by the New York offices. Wences seemed to already have a full grasp of the mechanics of Bitcoin. He talked about potential risks that only the best-informed Bitcoiners knew about and conversed knowledge-ably about monetary policy in the United States and the country where he had grown up: Argentina. When he spoke, it was in a gentle but candid way, giving the impression that much of what he said was a kind of personal confession.

"Bitcoin forced me to realize I didn't understand money," Wences liked to say.

Charlie and Erik couldn't immediately place him among the

familiar Bitcoin types. He wasn't a libertarian, raving about the transgressions of the government, and he wasn't a tech nerd, fascinated by the cryptography. When he left, after a polite conversation, Erik and Charlie still weren't sure why Wences had come.

At the time of his visit to New York, Wences was in the first year of running a startup, Lemon Digital Wallet, which provided a way for customers to keep all their credit cards and coupons in digital form on a smartphone. But this startup was only the latest in a career that had already turned him, at age forty, into one of the most successful technology entrepreneurs to ever come out of South America. In his teens, he had established Argentina's first Internet provider, and in his twenties he founded a company that became a sort of Latin American E*Trade. Backed by the storied New York investor Fred Wilson, the company made $750 million for its investors when it was sold to Banco Santander. Wences and his wife Belle used some of his new fortune to buy a catamaran and sail around the world with their young children. When they returned, the family moved to Silicon Valley.

Wences had first learned about Bitcoin in late 2011 from a friend back in Argentina who thought it might give Wences a quicker and cheaper way to send money back home. Wences's background in financial technology gave him a natural appreciation for the concept. After quietly watching and playing with it for some time, Wences gave $100,000 of his own money to two high-level hackers he knew in eastern Europe and asked them to do their best to hack the Bitcoin protocol. He was especially curious about whether they could counterfeit Bitcoins or spend the coins held in other people's wallets— the most damaging possible flaw. At the end of the summer, the hackers asked Wences for more time and money. Wences ended up giving them $150,000 more, sent in Bitcoins. In October they concluded that the basic Bitcoin protocol was unbreakable, even if some of the big companies holding Bitcoins were not.

By the time he visited the BitInstant offices, Wences had become a Bitcoin believer, and he was intent on spreading the idea among his powerful friends in Silicon Valley, a place that had so far largely ignored Bitcoin, but that would be vital if the technology was going to move into the mainstream.

For Wences, the allure of Bitcoin went deeper than just professional interest, to a time before he was wealthy and successful, during his childhood in a country that had been—and remained—locked in a seemingly intractable battle with its own currency.

There was rarely a time during Wences's youth when Argentina was not in some sort of financial crisis. In 1983, after years of staggering inflation, the government created a new peso, each one of which was worth 10,000 old pesos. That didn't work and so in 1985 the new peso was replaced by the austral, which was worth 1,000 new pesos. Seven years later, continuing inflation led the government to go back to the peso, but this time pegged to the dollar, an experiment that eventually ended with a crushing financial crisis. During most of this time, inflation ran at over 100 percent a year, meaning that the value of money in the bank fell by half each year and often much more than that.

Wences was descended from one of Argentina's aristocratic families, but his particular branch had lost everything and ended up on a rustic sheep ranch out in the emptiness of Patagonia. When his father delivered wool and the check didn't come through for a month, the value of the family income could fall sharply because of inflation, setting off yet another round of household cutbacks.

"I think I understand economics better than most people because I grew up in Argentina," he would say. "I've seen every single monetary experiment you can imagine. This is the street smart economics. Not the complex PhD economics."

One particular incident had seared itself into Wences's memory. In 1984, during the first major episode of hyperinflation after the Argentinian military junta lost power, Wences's mother came to get him and his two sisters from school. His mom was carrying two grocery bags filled with money—the salary she had just been given in cash. She rushed with Wences and his sisters to the grocery store and had them run through the aisles, grabbing as much food as possible before the hyperinflation caused the goods to be repriced. A man walked through the aisles all day doing nothing but repricing the items on the shelves to keep up with the rapidly changing value of the peso. When Wences and his mother got to the register, he and his sisters would run back and grab more food if they still had any money left. Holding on to money was equal to losing it.

These experiences gave Wences insights into the nature of money that most people in the world learn only from textbooks. In America, the dollar seamlessly serves the three functions of money: providing a medium of exchange, a unit for measuring the cost of goods, and an asset where value can be stored. In Argentina, on the other hand, while the peso was used as a medium of exchange—for daily purchases—no one used it as a store of value. Keeping savings in the peso was equivalent to throwing away money. So people exchanged any pesos they wanted to save for dollars, which kept their value better than the peso. Because the peso was so volatile, people usually remembered prices in dollars, which provided a more reliable unit of measure over time.

As Wences avidly pored over all the available writing about Bitcoin in the first months after discovering it, it seemed clear to him that for people in places like Argentina, Bitcoin might provide a much more efficient place to store money than the dollar. In Argentina, dollars had to be purchased through shady money changers, and were saved in closets or under the mattress. The promise of a virtual currency that could be bought and stored online, accessed from anywhere, and secured with a private key looked like a significant improvement.

Wences began by purchasing a growing stockpile of Bitcoins from Mt. Gox in early 2012 and joined in the conversation on the forums and chat channels. When he wasn't playing with Bitcoin, he devoured several books on the history of money, most significantly *Debt: The First 5,000 Years*, a cult favorite in the Occupy Wall Street movement and in certain transgressive corners of Wall Street. The book, by anthropologist David Graeber, argued that historians and economists have wrongly assumed that money grew out of barter. In fact, Graeber argued, and Wences came to believe, barter was never common and money was actually an evolution of credit—a way of tracking what people owed to each other. People used to just keep a mental tally of what they owed each other, but money provided a way to expand the system more broadly among people who didn't know each other.

As he read, Wences felt that after twenty years of working on financial technology, he was finally coming to understand money for the first time. He saw that Bitcoin's lack of any apparent intrinsic value didn't matter when looked at against the history of money. The reason gold itself had been used as money was not that it was valuable; it had become valuable because it was used as money. And it was used as money because it did what all good money did: it served as a sort of physical ledger on which society could keep track of who was owed what. Each piece of gold represented a slot on the ledger of all outstanding gold, which anyone could verify by checking the mass and volume of the gold.

"We don't use gold because it's pretty—that was a stupid assumption of mine and many other people," Wences would tell anyone who would listen during these days when he was totally immersed in the history. "No, we use it in jewelry because it's very expensive. It's not expensive because we use it in jewelry."

"What is the value?" he would ask. "It's that it is the ledger. You put the ledger on your neck to show power and wealth. It isn't a ledger where you have to trust a bank or anyone else."

Bitcoin, Wences came to believe, was a purer version of that sort of ledger—a commonly verifiable place where everyone could keep track of who owned what.

Despite his fervor, Wences initially had trouble drumming up much interest among his Silicon Valley friends beyond a few fellow South Americans, who had grown up in places with similarly screwed-up currencies. Mostly he just got skeptical looks. For those who had heard of it, the first question was usually about whether it was anything more than a token for online drug dealers. Some remembered David Chaum's DigiCash back in the 1990s, but anyone familiar with that experiment knew that it had gone under. The bigger question was why something like this was necessary in the first place. Credit cards and $20 bills did everything that most of Wences's friends needed when it came to spending money. Why should they trust a digital code that had nothing backing it but the computers of some libertarian nerds?

After months of trying, Wences finally made a breakthrough with one of his best friends, and the one whose opinion in this area mattered most: David Marcus, who had recently become the president of PayPal. Like Wences, Marcus was a foreigner in the Valley. He had grown up in France and Switzerland, and had the same slender stature, unassuming presence, and seemingly effortless sophistication as Wences. But after spending a decade on payment startups, Marcus was used to hearing grand claims about technologies that would revolutionize the way money moved around the Internet. He also had experienced the overbearing regulatory scrutiny that falls on any company that wants to deal with money.

But in the fall of 2012 Marcus had a conversion moment when the Argentinian government ordered his company, PayPal, to cut off direct payments between Argentinians, a new prong in the government's effort to slow the movement of pesos into other currencies. With Wences's arguments ringing in his head, Marcus watched as the policy went into effect and the price of Bitcoin rose, suggesting to Marcus that Argentinians were seeking out Bitcoin as a way around the government's restrictions. He quietly set up an account with Mt. Gox and began buying coins. In doing so, Marcus became one of the first of many important converts that Wences would win to the Bitcoin cause.

WENCES RAN HIS digital-wallet company with an old friend in Argentina, Federico Murrone, or Fede. Unlike Wences, who had an aristocratic lineage, Fede came from a working-class family and looked like a tough biker. The two had connected as teenagers on Wences's first startup, creating Argentina's first Internet provider, and they had been close friends ever since, with Fede providing the programming smarts for Wences's ideas, always from Argentina, where Fede stayed to be close to his family.

Wences traveled to Buenos Aires every few months to check in with Fede and his team of Argentinian coders. Each time Wences visited in 2012, the reminders of what it was like to live in a country with broken money strengthened his belief in the potential for Bitcoin.

Like other smart visitors to the country, Wences went to a black market money changer whenever he needed pesos to spend. Credit cards and ATMs were available, but they provided pesos at the official government exchange rate, which was about 35 percent lower than the rate available on the street in 2012. The government wanted to make changing money into dollars unattractive, with its official exchange rate, because it was afraid that its citizens would sell off all their pesos for dollars, driving the exchange rate down even further and devastating the economy. The government had recently started fining economists who challenged its official exchange rate. As 2012 went on, the situation grew progressively worse, and this is what had led the government to crack down on PayPal.

The inflation rate wasn't the only problem with the local financial system. As in many developing countries, it was incredibly hard to open a bank account and even harder to get a credit card. Despite having grown up in Argentina, Wences had never had an Argentinian bank account. People were left to pay their bills in cash at the drugstore, so they had to carry around wads of 100-peso bills. This too, seemed like something that Bitcoin, with its secure digital wallet, could help address.

At the Lemon offices in Buenos Aires, Wences and Fede were supposed to be working on their new startup, but they would end up spending hours playing with Bitcoin and talking about how they might harness its potential. In late 2012 the two men organized the first-ever Bitcoin Meetup in Argentina at a favorite whiskey bar. It was sparsely attended, other than by the friends that Wences and Fede had already sold on the technology. The small crowd was not surprising given how hard it was to get Bitcoins in Argentina. It was incredibly expensive and difficult to transfer money from an Argentinian bank to Mt. Gox or another foreign Bitcoin exchange. And no Argentinian bank would work with a domestic Bitcoin company.

But there was a budding conversation about Bitcoin on an Argentinian website dedicated to protecting online freedom. When Wences was in Argentina, he would offer to sell some of his Bitcoins after work at a bar near the Lemon offices. Each time, a different crew of people would show up, but one older gentleman kept coming back, buying a little more each time. He had a silent, sullen countenance and didn't seem technologically sophisticated.

After the man made a particularly large purchase one day, Wences gently asked him if he understood the risks involved with Bitcoin.

"It seems to me like this is a lot of money, and this is very risky," Wences told him as politely as he could. "You know you could lose it all?"

"How many times has your family lost everything keeping their money in the peso?" the man asked Wences.

"Three, maybe four, times," Wences said.

"Yes. For me it's been more times than that," the man said.

The man admitted that he had the option of putting his money in dollars but that this would require him to take a distorted exchange rate and then hide the bills in his closet. And who knew when the dollar might suffer the same problems as the peso?

"There is no way you can convince me to keep my money in the peso," he said.

BITCOIN HAD CAUGHT Wences at a decisive moment in his life—what an American might call a midlife crisis. He already had many successes under his belt, as was evident from his estate in the rolling hills above Palo Alto, with two homes, a swimming pool, tennis courts, and views down to the bay. In addition to the tens of millions of dollars Wences had earned from selling past startups, he had been surprised to discover that he also had a knack for picking winning investments in his friends' companies.

But he had recently been hitting up against failure for the first time. Lemon, his current startup, had grown out of the decline of his previous startup, Bling Nation, and many of Wences's friends wondered whether Lemon was the result of the kind of passion necessary to succeed in Silicon Valley or was just Wences's attempt to prove that the failure of Bling Nation had been an anomaly. There were already signs that Lemon was not getting the kind of pickup that Wences had imagined. And, as with all startups, it required more time from its chief executive than any one person had in a day. This was Wences's twelfth startup, depending on how you counted, and his wife once again felt like a single mother for their three children. Wences and Belle had already agreed that Lemon would be his last startup.

These difficulties played into larger insecurities that Wences had managed to sweep under the carpet until now. For all the money his past startups had made him, none had quite achieved their grand original goals. Back in Argentina, he had hoped that his first company, Patagon, would provide a way to extend financial services to the hundreds of millions of South Americans without a basic bank. In the end, though, he couldn't get a banking license, and the online financial firm he created was used mainly by South America's wealthy elite.

For Wences, Bitcoin seemed to address many of the problems that he'd long wanted to solve, providing a financial account that could be opened anywhere, by anyone, without requiring permission from any authority. He also saw an infant technology that he believed he could help grow to dimensions greater than anything he had previously achieved.

Wences's wife, Belle, was used to watching Wences dive head-first into new technological discoveries. His easily incited passion and his ability to convey it were part of what made him such a great startup salesman. He could impart his excitement with a rare skill. But usually the initial ardor passed before long. That was showing no signs of happening with Bitcoin. As 2012 went on, Belle realized that this might be different from his previous endeavors. Belle herself resented how much time Bitcoin was taking out of Wences's already full schedule. But even she was becoming entranced by the almost mythical nature of this currency and its mysterious founder. She soon started swapping her own theories with Wences on the identity of Satoshi.

IN EARLY JANUARY, Wences traveled with a group of some of the West Coast's most wealthy and powerful men to an isolated lodge in the Canadian Rockies with its own wine cellar, sauna, and private staff. Their host was Pete Briger, whom Wences had met a few years earlier through an organization for young chief executives. Even among the rich and powerful, Briger stood out. After attending Princeton and working at Goldman Sachs for fifteen years, he had risen to the top of the Fortress Investment Group, a firm overseeing an array of enormous private equity and hedge funds.

Briger was a big gruff man, who was known for his bold bets on distressed debt—the troubled bonds and loans that everyone else was too afraid to touch, and that gave Briger and his firm arm-twisting leverage over large companies and occasionally small countries. He sometimes called himself a "financial garbage collector" and he looked the part. In 2009 Briger had been named cochairman of Fortress, which then controlled investments worth around $30 billion, including the resort company that owned the lodge where the men were staying.

Wences was not an alpha male like most of the other guests. He liked to stay in touch with his humble origins in Patagonia, and his driveway was filled with Subarus instead of Teslas or sport cars. Rather than taking luxury vacations, Wences used his time off to go with his wife to Burning Man, and he had recently done a vision quest—involving days without any creature comforts—in the wilderness of the Andes with one of his best friends from his younger years in Argentina. But Wences had a good-natured self-confidence and a willingness to listen that had always allowed him to get along easily with hard-driving power players.

The morning after they arrived at the Valemont lodge, Wences, Briger, and the rest of the men climbed into a red-and-white Bell 212 helicopter sitting just outside the lodge and lifted off toward the high white peaks, for a day of heli-skiing. In the afternoon, the group returned to the lodge and sat around in the expansive common room, an enormous fire crackling away. This was not a crowd to chat about kids and the upcoming Super Bowl. The men had dedicated their lives to making money and Pete pressed them to present their best investment ideas.

"Pete, I told you, I'm interested in Bitcoin," Wences said when his turn came to talk. "It hasn't changed."

Wences drew the group in with an explanation of the basic notion of a new kind of network that could allow people to move money anywhere in the world, instantaneously—something that these financiers, who were frequently moving millions between banks in different countries, could surely appreciate.

"You can call someone in Jakarta on Skype," Wences told them. "You can see them and you can hear them and there's a synchronous connection with a lot of bandwidth. There's a ton of magic happening, which is incredible. And you hang up and you want to send them one cent and that's not possible. That's ridiculous. It should be a lot easier to send a cent than to see video and audio."

The blockchain technology made that previously impossible task possible. But it was much more than that, Wences emphasized. It was the next step in the evolution of money. He tried to explain his recent discoveries about the ledger as the foundation of all money. With Bitcoins, unlike pesos or dollars, everyone using them knew exactly how many existed, and they were not tied to one country. Unlike gold, which was universal but difficult to acquire and hold, Bitcoins could be bought, held, and transferred by anyone with an Internet connection, with the click of a mouse.

"Bitcoin is the first time in five thousand years that we have something better than gold," he said. "And it's not a little bit better, it's significantly better. It's much more scarce. More divisible, more durable. It's much more transportable. It's just simply better."

Pete had a habit of taking long, anxiety-inducing pauses before responding to anything, and his first questions for Wences were distinctly skeptical. But his subsequent questions suggested that something was clicking. Pete's job as an investor in distressed companies made him good at spotting broken systems, and the more he thought about it, the more broken the current methods of moving money around the world seemed to him.

Something else caught Pete's attention. Wences had put his wallet where his mouth was. Throughout 2012 Wences had methodically ramped up the pace of his Bitcoin purchases, so that now he had over 10 percent of his net wealth—tens of millions of dollars—in Bitcoin. Pete respected numbers and bold, confident moves like the one Wences had made.

From the ski lodge, Pete e-mailed one of his most trusted lieutenants at Fortress, Bill Tanona, and asked what Bill knew about Bitcoin. When he got back to San Francisco he opened a Mt. Gox account and quickly built up his own $100,000 position in Bitcoin. At work, he started talking with Tanona and a few other colleagues about how Fortress could get involved in this new market.

●

CHAPTER 16

December 2012

For all the new mainstream interest, the most successful entrepreneur in the Bitcoin world was still Ross Ulbricht, the operator of the world's largest online drug bazaar. Silk Road had continued adding new members and new products through 2012. Some $1.2 million worth of Bitcoin was changing hands each month, spinning off $92,000 in commissions for Ross. By the end of 2012 there were seventy thousand different topics on Silk Road's forum, and there were even resident security experts who helped users ensure their anonymity and a resident doctor who answered questions about drugs and their health effects.

Initially, Ross had enjoyed the success by traveling to Southeast Asia and Costa Rica. But as the year went on, the site increasingly required all-consuming work. Ross now had several moderators and administrators on staff who helped him deal with customer support and mediate disputed transactions. He chose members whom he trusted, even when he didn't know their identity.

In the fall of 2012 Ross had moved in with a childhood friend on a hilly street in one of San Francisco's residential neighborhoods.

He could have afforded his own place, but by now he was trying to leave as few traces as he could for the authorities to pick up on. His work on Silk Road was done at an Internet café around the corner from his friend's house; at this café he would log in remotely to Silk Road's servers, making it that much harder for anyone to find him.

Ross was becoming acutely aware of just how difficult it was to remain anonymous even with the best technologies. Over the summer, a Silk Road user had managed to follow a series of transactions and find one of Silk Road's main Bitcoin wallets, which contained coins worth about $2 million. This didn't cover any losses, but it was a reminder that while Bitcoin did not require users to provide an identity, accounts were pseudonymous, attached to a particular identity, rather than anonymous. In Australia, police traced transactions to make the first arrests of Silk Road vendors in that country.

None of this, though, dented Ross's boldness and ambitions for the site—if anything, he grew more committed as time went on. On the forums, under his screen name Dread Pirate Roberts, or DPR, he wrote that he would keep this up to his "last breath":

> Once you've seen what's possible, how can you do otherwise? How can you plug yourself into the tax eating, life sucking, violent, sadistic, war mongering, oppressive machine ever again? How can you kneel when you've felt the power of your own legs?

As Dread Pirate Roberts, Ross became a kind of folk hero for his members, engaging with them on the Philosophy, Economics, and Law section of the forum and later on DPR's Book Club, where he advocated for a world in which "the human spirit flourishes, unbridled, wild and free!"

As time went on, though, it was hard to avoid the growing reminders of the dangers of living in an anonymous world with no source of authority. In November 2012 a hacker threatened to release an enormous trove of data about Silk Road users if Ross didn't pay a ransom. That was soon followed by a denial-of-service attack that eventually forced the site down. The only way Ross was able to get the attack to stop was by paying the attacker $25,000. When the site came back online, Dread Pirate Roberts's style and approach had shifted, leading some users to suspect that the site had changed hands. Ross explained that he was changing his writing style to elude capture.

In November, Ross flew to Dominica, an island in the Caribbean known for being an easy place to secure "economic citizenship" (Roger Ver was also trying to obtain citizenship from the country). The small island offered a passport in exchange for a $75,000 donation. The sum was no problem for Ross and he began filling out the application on his laptop, listing his profession as "IT consulting." A new passport would allow him to move that much further out of the reach of a government that he knew was chasing him.

He was, though, getting used to his new life. When he chatted with a Silk Road member, scout, who was thinking about joining his staff, Ross answered scout's concerns about getting arrested by explaining why he believed it would be hard to ever get caught.

"put yourself in the shoes of a prosecutor trying to build a case against you," he said in a chat with scout. "Realistically, the only way for them to prove anything would be for them to watch you log in and do your work."

But Ross acknowledged how much even the small possibility weighed on him.

"the biggest con about this work is not the risk of going to jail or having your life disrupted," he wrote; "it's getting used to and living with that possibility no matter how remote."

"and," he added, "keeping your work a secret."

By now he had been hardened enough that he knew how to keep things to himself. Even the friend he was living with and the girl he began dating didn't know. The only people with whom he could be honest were the users and administrators of his site, who didn't know his identity, and it was becoming increasingly hard to believe that he could trust even them. Silk Road forums were rife with debate about which users and vendors on the site were likely to be undercover cops. One of the most vigorous debates sprang up around a user named nob. Toward the end of 2012, nob put up a listing for a kilogram of cocaine for $27,000 in Bitcoins. nob had done almost no reviewed sales of drugs on the site and many other users were very suspicious.

"If this acct isn't [law enforcement], it's some other bullshit for sure," a user named MC Haberdasher wrote on the forum. "I'd rather wake up from a heroin induced blackout sitting bitch in a car full of fat chicks listening to speed garage than even attempt to order from this guy."

In this case, though, Ross trusted nob, who had slowly built a relationship with him over the course of the previous year. Ross decided to help nob sell his kilogram of cocaine, connecting him with one of the site administrators, chronicpain, who had been the first employee Ross hired back in 2011. The administrator was, in real life, Curtis Green, a forty-seven-year-old poker player and grandfather living just outside Salt Lake City.

Green found a buyer for nob's cocaine and offered to receive the package at his home before sending it on to the buyer. The package was delivered to Green's house on January 17, 2013. Just as he took it inside and was opening the package to check its contents, a SWAT team swarmed in. As the agents spread through the house, they found a stack of black, custom-made Bitcoin-mining machines. The floor around the computers, and in the rest of the house, was littered with hardened dog shit.

Even after Ross learned about Green's arrest—and his release on bail—he did not assume that it was nob who had compromised the deal. Ross had always been somewhat skeptical about Green, believing that he was doing it for the money rather than the ideals. Ross asked nob (who he still believed was a powerful drug dealer) if he could have Green "beat up, then forced to send the Bitcoins he stole back."

nob agreed to the proposition. But a day later, Ross changed his mind: "can you change the order to execute rather than torture?"

Ross explained to nob that he was concerned that Green would give the authorities information about Silk Road users, potentially jeopardizing the whole site and its grand mission. He said that he had "never killed a man or had one killed before, but it is the right move in this case."

The federal agents who had Green in custody, and who were the undercover puppeteers behind the user nob, obliged by staging Green's death (without actually killing him), and e-mailing bloody photos to Ross. When the photos came through, Ross responded that he was "a little disturbed, but I'm ok."

"I'm new to this kind of thing is all," he said, before quickly adding: "I don't think I've done the wrong thing."

The purported murder of Green was paid for with a transfer of $80,000 to a Capital One Bank account in Washington, DC. The money was sent through an anonymous money-transferring service in Australia that hid the location and identity of the sender. But the agents were already digging into the wealth of information on Green's computers, seeking clues to find their way to their real quarry, Dread Pirate Roberts himself.

CHAPTER 17

●

January 2013

Ross Ulbricht was not the only Bitcoin entrepreneur who had gotten himself into something bigger than he could have ever imagined. In January Charlie Shrem's BitInstant was taking in over $250,000 in commissions each month on record transaction volumes.

But the growth obscured strains that were threatening to tear Charlie's company apart. The fights with David Azar that had started almost as soon as BitInstant took David's investment had grown worse and usually ended in a shouting match or a slammed-down phone. In December, Charlie and Erik Voorhees had looked to David's investment partners, the Winklevoss twins, to help foster a more productive relationship.

The brothers had been relatively hands-off after putting in their $550,000. But they had grown concerned from afar. The e-mail chains between Charlie and David signaled that the twins were not dealing with the cool, calculating entrepreneurs of their Harvard alumni circles. They saw that Charlie's initially attractive energy came with a distressing inability to concentrate on one task. Between constant travel and media appearances, Charlie was relishing, perhaps too much, the elevated social status that Bitcoin was giving him. When Charlie did talk business, he often seemed more intent on selling the idea of Bitcoin than of his own company.

There was another more immediate problem that the twins hadn't bargained for. Earlier in the year, Erik and a friend he had brought into BitInstant, Ira Miller, had started an independent company called Satoshi Ltd. with a number of subsidiaries. One was a technology called Coinapult that BitInstant used to send Bitcoins via e-mail. Another, Paysius, allowed merchants to accept virtual currencies.

The Winklevoss twins asked how Erik and Ira could run those businesses at the same time that they were working full-time for BitInstant. Erik and Ira proposed solving the issue by merging Satoshi Ltd. with BitInstant in exchange for a higher equity stake in BitInstant—all that David and the twins had to do was give up 1.5 percent of their own stake in the company.

Around the New Year, Erik wrote up a lengthy strategy document listing how a merger could be handled, allowing the company to go after new markets like mobile payments in Africa and poker sites in need of payment networks around the world. The document reflected the team's big ambitions. Erik and Ira didn't want Bit-Instant to be just a place to buy Bitcoins. They wanted to offer all the services that banks did, in a new, cheaper, and more democratic way.

But the Winklevoss twins and David Azar were thinking in more immediate and practical terms. Glancing at the pages of long-term strategy, they blanched at the value that Erik and Ira assigned to Satoshi Ltd.

The twins wrote increasingly peeved e-mails to Charlie, pushing him to resolve the situation without giving in to Erik and Ira. The conversations between the twins and Charlie began to end with the same sort of recriminations that had been so common between Charlie and David weeks earlier. Charlie and his team appeared to the twins like inexperienced entrepreneurs who didn't know how to put business interests above social and political allegiances. The Winklevoss twins, meanwhile, confirmed the fears of the BitInstant team regarding what happened when people who didn't care about the big principles underlying Bitcoin tried to make money in the space.

Charlie and Erik reached out to Roger Ver, Charlie's first investor, hoping he might be able to resolve things from Tokyo. Their idea was that Roger could buy out the stake that the twins and David had taken in BitInstant.

"My one hope was that perhaps the Winklevii would be far more helpful and productive, but a long insult-filled call between Cameron and Charlie today proved that my hope was naive," Erik wrote to Roger in early January.

Charlie and Erik wrote a lengthy, acerbic letter to the twins, pleading for a resolution that would allow both sides to go their separate ways.

"If we're all being honest, then it's clear we neither need nor want your money, and you neither need nor want to be risking your money with a team that you believe to be childish and 2/3rds expendable," the letter said. "Let's be gentlemen and move on. If you are so interested in building a Bitcoin business, and you are so skillful at navigating these waters, then I welcome you to go and do it."

The twins considered selling to Roger. But they also believed BitInstant was a good idea that could work under the right management. In January BitInstant had its best month ever, processing almost $5 million in transactions. The price of a Bitcoin, meanwhile, had risen from $13 at the beginning of the month to around $18 at its end. Some of this was due to the twins themselves. They had asked Charlie to continue buying them coins with the goal of owning 1 percent of all the Bitcoins in the world, or some $2 million worth at the time. This ambition underscored their commitment to sticking it out with Bitcoin.

The tension came to a breaking point at the end of January. Patrick Murck, the general counsel at the Bitcoin Foundation, flew in from Seattle to see if he could help Charlie and Erik make their argument to the twins. In a meeting in the BitInstant conference room, Charlie, Erik, and Patrick, sitting on one side of the table, offered to provide Maguire Ventures, the entity put together by David and the twins, with a full refund for the money they had put in. The twins responded angrily that they would accept no less than five times what they had put in. They also said that the technology being offered by Erik's company, Satoshi Ltd., was worth little. Erik and Ira responded by walking out of the room as the twins "continued with emotional insults and absurdities," Erik wrote in an e-mail after the meeting.

The next day Erik and Ira sent in their resignations and moved into the offices of Larry Lenihan and FirstMark Capital; Lenihan had always been more interested in investing in Erik than in Charlie.

Charlie, Roger, and Erik were in constant conversation, contemplating whether Charlie should join Erik, and if the whole group should sue the Winklevoss twins. They ultimately decided not to sue—mindful of the way the twins had responded when Mark Zuckerberg left them out of Facebook.

Charlie decided he couldn't leave the company he created, but when he went to work the next day, he did not go in peace. He demanded that Maguire Ventures deliver the final installment of the investment it had agreed to make the previous fall:

"You guys are screwing up my company, and Ira and Erik left because of it. Give me my money or I will wire it all back to you today."

Roger, who still had a 15 percent stake in the company, continued pushing the twins to sell their stake in the company or let Roger sell his:

You guys obviously don't understand Bitcoin, or BitInstant.

You are destroying your equity and mine, and I don't want to be any part of it.

If you disagree, then make me an offer for my 15% of BitInstant.

Name your price.

I will gladly sell it to you for less than the valuation you bought in at.

There was some confirmation of Roger's assessment a few days after Erik left, when Charlie got a letter from the latest bank to decide that it would no longer service BitInstant's accounts. It was unclear if BitInstant would have anywhere to put all the money customers were sending it. As the value of Bitcoin continued to shoot up, the value of Charlie's idea seemed to be falling apart before his eyes.

CHAPTER 18

●

February 2013

The desk where Wences Casares worked on his digital wallet, Lemon, was mounted on a treadmill, in an office overlooking the main shopping street in Palo Alto. His monitor was perched on a short pile of books, hardcover copies of *Debt: The First 5,000 Years*. When he spoke about Bitcoin with visitors to the office and invariably began talking about the history of money, he would frequently give them a copy of the book.

Wences shared the space with Micky Malka, an old Venezuelan friend and business partner. Micky was a big investor in Lemon and chairman of the company's board. Wences was, for his part, one of the largest investors in Micky's venture capital firm, Ribbit Capital.

Micky's recently opened fund was technically focused on financial services. But after Wences got Micky excited about Bitcoin, Micky was trying to find virtual-currency investments. Because there were so few viable Bitcoin companies around, Micky made the somewhat controversial decision to use his investors' money to buy Bitcoins themselves.

Both Micky and Wences turned the office into a kind of virtual-currency salon, hosting a constant parade of interested visitors. Among them was Pete Briger, the chairman of Fortress Investment Group, who dropped by soon after the skiing trip, with his deputy Bill Tanona. Wences marveled at how quickly Pete had managed to get others at Fortress excited about Bitcoin, but when he heard Pete speak about it he understood why. Pete, a normally reserved man, got fired up when talking about the inefficient "oligopoly" that the big banks had over money movement and the transaction fees that the oligopoly forced everyone else to pay. Wences was getting more of a response from Fortress—a Wall Street giant managing nearly $60 billion—than he was from Silicon Valley venture-capital firms with just a few hundred million dollars. Pete assigned Tanona to the almost full-time job of exploring potential Bitcoin investments, and also drew in another top Fortress official, Mike Novogratz. All of them began buying coins in quantities that were small for them, but that represented significant upward pressure within the still immature Bitcoin ecosystem.

The purchases being made by Fortress—and by Micky's team at Ribbit—were supplemented by those being made by the Winklevoss twins, who were still trying to buy up 1 percent of all the outstanding Bitcoins. Together, these purchases helped maintain the sharp upward trajectory of Bitcoin's price, which rose 70 percent in February after the 50 percent jump in January. On the evening of February 27 the price finally edged above the long-standing record of $32 that had been set in the hysterical days before the June 2011 crash at Mt. Gox.

ON THE AFTERNOON of Sunday, March 3, Wences boarded a Gulfstream two-engine jet at a private airport in San Jose favored by the Silicon Valley elite.

Wences was headed to one of the most exclusive, and secretive, annual gatherings of tech-industry power players, held at the Ritz-Carlton resort outside Tucson, Arizona, and hosted by the investment bank Allen & Co. Only a few hundred people were invited and it was private enough that the news media rarely even found out it was happening.

Wences flew to the conference on eBay's private jet. eBay owned PayPal, the company headed up by Wences's good friend David Marcus, and David was among the twelve passengers on the flight. He had been quietly working to make sure PayPal was ahead of the curve on virtual currencies and had pulled together a group in-house to look at how PayPal might harness the Bitcoin technology. He had also begun to talk about it with his boss, John Donahoe, the chief executive of eBay.

When the eBay jet touched down north of Tucson, the passengers were quickly whisked away in SUVs to the Dove Mountain Resort, which sat in the foothills of the mountains that separate Tucson and Phoenix. That evening, everyone congregated for drinks on the Tortelita Terrace and then proceeded to dinner on an immaculately maintained lawn overlooking the scrubby mountains.

This was the most casual dinner of the three-day event, with unassigned seating and a buffet to accommodate the guests arriving at uneven intervals. Wences took notice as the big names showed their faces: Twitter's chief executive, Dick Costolo; LinkedIn's founder Reid Hoffman; Rupert Murdoch's son, James; and perhaps the most recognizable venture capitalist in Silicon Valley, Marc Andreessen, an enormous man with a shiny bald head.

Wences found his way to a table with another budding Bitcoin nut, Chris Dixon, one of the up-and-coming stars at Andreessen's firm, Andreessen Horowitz. The men quickly began comparing ideas. Dixon explained that he had gotten excited about the importance of the blockchain protocol as a new way of moving value around the world, just as the Internet protocol had provided a decentralized way to move information. Dixon had been pushed to think about this by the writings of Fred Wilson, the New York venture capitalist who had backed Wences's first big company.

Wences smiled with gratitude to find someone who had seen the beauty of the system without his help. Wences, in turn, told Dixon about the international potential he saw for Bitcoin, in countries like Argentina where people lack a safe place to keep their money. Dixon hadn't thought much about that opportunity and asked Wences to tell him more.

They were interrupted by Henry Blodget, a former Wall Street analyst and founder of the news site *Business Insider*, who asked what they were talking about: he had never heard of Bitcoin. Wences responded with his favorite introductory line: "It's the best form of money the world has ever seen."

Blodget's famously childlike curiosity provided a great opening for Wences to work through all of his finely honed arguments.

After touching on the history of money and Bitcoin's advantages over gold, Wences explained his back-of-the-envelope calculations of what Bitcoin might be worth if people began to realize its value as a substitute for gold. All the gold in the world was worth around $7 trillion. If Bitcoin became even half as popular, that would put the value of each Bitcoin at around half a million dollars—or about fourteen thousand times more than its $34 value that day in March.

The conversation continued as the sun went down and the desert air grew chilly. The little crowd around Wences's table grew, with Marcus and others stopping by.

Wences saw the interest build when he told one of his newest stories from Argentina. A friend of his sister had recently wanted to buy an upscale $1.5 million apartment in Buenos Aires. As with most Argentinian real estate transactions, the seller—distrustful of the peso—wanted the payment in dollars and in cash, no small feat when the sum was $1.5 million. The bigger problem was that the sister's friend, like many wealthy Argentinians, kept his savings in dollars in an American bank account. To transfer the money into an Argentinian bank and then take it out in cash would eat about 10 percent of the money in bank and exchange fees—some $150,000—and would involve several days of waiting. To get around this, the sister's friend purchased $1.5 million worth of Bitcoins from Mt. Gox. Once the friend had the coins, he took his Bitcoin wallet to the signing for the apartment in Buenos Aires and transferred it over to the seller, with the notary as witness. Afterward, Wences's sister sent him a picture of the two old men holding up their smartphones and smiling.

To prove how easy this all was, Wences asked Blodget to take out his phone and helped him create an empty Bitcoin wallet. Once it was up, and Wences had Blodget's new Bitcoin address, Wences used the wallet on his own phone to send Blodget $250,000, or some 6,400 Bitcoins. The money was then passed to the phones of other people around the table once they had set up wallets. Anyone could have run off with Wences's $250,000, but that wasn't a risk with this particular crowd. Instead, as the money went around, Wences saw the guests' laughter and wide-eyed amazement at what they were watching.

The next two days were filled with panels covering topics like "eBay and Innovation" and "China: The Road Ahead." In the afternoon there were scheduled activities: tennis, horseback riding, and clay-pigeon shooting, among others. During the interludes Wences was approached constantly by people who had heard the Sunday evening conversation or heard about it. LinkedIn founder Reid Hoffman pulled Wences aside to ask more, as did Michael Ovitz, the former president of Disney. During a hike on Wednesday afternoon, Wences spent the entire time explaining the concept to Charlie Songhurst, the chief of strategy at Microsoft. At night, many of the same people approached David Marcus. As the president of PayPal, he would have as informed a view as anybody on the viability of Bitcoin.

"What do you think of this?" they asked him. "Is this real?"

Marcus replied that he already believed in the idea enough to put his own money into it. They shouldn't invest money they couldn't afford to lose, he said, but it was certainly worth some investment.

On Monday, the first full day of the conference, the price of Bitcoin jumped by more than two dollars, to $36, and on Tuesday it rose by more than four dollars—its sharpest rise in months—to over $40. On Wednesday, when everyone flew home, Blodget put up a glowing item on his heavily read website, *Business Insider*, mentioning what he'd witnessed (though not specifying where exactly he'd been, or whom he'd talked to):

> I was at a technology conference earlier this week, and the most popular topic of casual conversation was Bitcoin, the electronic currency invented and unleashed a few years ago.
>
> One of the things that's most fascinating about Bitcoin, I have learned, is that it entrances fanatical conspiracy theorists, clear-eyed pragmatists, and diehard skeptics alike.

Songhurst, the Microsoft head of strategy, who had learned about Bitcoin during his hike with Wences, wrote up a paper and circulated it among some of the most powerful investors in Silicon Valley, channeling Wences's arguments:

> We foresee a real possibility that all currencies go digital and competition eliminates all currencies from non-effective governments. The power of friction-free

transactions over the internet will unleash the typical forces of consolidation and globalization and we will end up with six digital currencies: US Dollar, Euro, Yen, Pound, Renminbi, and Bitcoin.

The question then becomes, is Bitcoin viable if the government digital ledger systems are just as good? We think yes, for two reasons:

1. There will always be transactions for which "official money" is less good than Bitcoin
2. If you live outside the US, it is dangerous to have all your money controlled by a state where you have no rights.

In three days, Wences had reached more powerful people than Bitcoin had in its previous four years of existence.

DESPITE THE SURGE of excitement, the interest Wences was encountering was still far from uniformly positive. More than a few people in Arizona left unconvinced that the technology would work and survive government scrutiny. Much of this skepticism had the same root as the excitement, and that was Silicon Valley's defining, and cautionary, experience with financial technology: PayPal.

PayPal, of course, still existed, owned by eBay and run by Wences's friend David Marcus. But what made people wary was not the current incarnation of PayPal, but instead the company's early days, when it had ambitions to be something much bigger.

PayPal had been founded back in 1998 by Peter Thiel and Max Levchin, among others. Thiel was an avid libertarian, who had wanted to use Levchin's cryptographic expertise to fulfill the Cypherpunks' dream of sending money through encrypted channels, between private individuals and in particular between mobile devices like the PalmPilots of that time. In early staff meetings, Thiel gave speeches that could almost have come from the Cypherpunk mailing list.

"PayPal will give citizens worldwide more direct control over their currencies than they ever had before," he said.

PayPal grew quickly, but in 2001, as the company readied for an initial public offering, it hit roadblock after roadblock from lawmakers concerned about the possibilities for money laundering and other illegal activities. New York Attorney General Elliot Spitzer said PayPal was breaking the law by facilitating payments for gambling companies, and the Department of Justice decided PayPal was violating the USA Patriot Act. The new limits and restrictions imposed took it further and further from its ambitious original goals. Thiel and Levchin left PayPal soon afterward.

This had scared much of Silicon Valley away from tinkering with finance, which was seen as largely resistant to new technology because of all the regulations. But the PayPal experience also explained why there was a hunger for the idea of a virtual currency. There was a lingering memory of this unfulfilled dream of Silicon Valley. While the Internet had freed information and communication from the postal service and the publishing industry, the Internet had essentially never disrupted money, and dollars remained bound by the old networks run by the credit card companies and the banks.

In the month before the Arizona conference, Thiel himself had been poking around in the virtual-currency space once again, looking for projects that might take advantage of the blockchain, without getting too bound up in a currency that could piss off government officials. Chris Dixon, Wences's conversation partner at that Arizona dinner, had also been agitating to get his firm, Andreessen Horowitz, to look at cryptocurrency startups and had been finding a receptive ear in his boss, Marc Andreessen.

They had both found their way to the new company being created by Jed McCaleb, the original founder of Mt. Gox. Jed's new company, named Ripple, was a cryptographic network that could be used to send any currency, not just Bitcoins. That made it less threatening to governments and banks and more attractive to people like Andreessen and Thiel, who both offered small seed investments.

But both of these key Silicon Valley figures were also getting more comfortable with Bitcoin itself. The investment firm that Thiel had helped create with some of his PayPal riches, the Founders Fund, began talking with an engineer at Facebook who had founded an e-mail list for Silicon Valley insiders, dedicated to Bitcoin, about joining the firm to look for virtual currency investments.

The growing openness to Bitcoin was helped along by Silicon Valley's ballooning sense of self-importance in early 2013. With the Nasdaq composite stock index soaring, shares of Google at an all-time high, and startups selling for mind-boggling sums, many in the tech industry believed that they were going to be able to revolutionize and improve every element of modern life. Investors and entrepreneurs were cooking up ever more ambitious schemes involving virtual reality, drones, and artificial intelligence, alongside more quotidian projects, like remaking public transportation and the hotel industry. The PayPal founders were among the most ambitious, with Thiel advocating for floating structures where people could live outside the jurisdiction of any national government. Elon Musk, an early PayPal employee and founder of SpaceX, was aiming for the colonization of Mars. If there was ever a time that Silicon Valley believed it could revive the long-deferred dream of reinventing money, this was it. A virtual currency that rose above national borders fitted right in with an industry that saw itself destined to change the face of everyday life.

●

CHAPTER 19

March 2013

At the same time that Bitcoin's reputation was getting a makeover in Silicon Valley, the physical infrastructure of the Bitcoin network was also undergoing an extensive transformation.

For much of the previous year and a half, the computing power underpinning the network had grown steadily, but slowly. Over the course of 2012 the amount of computing power on the Bitcoin network barely doubled. What's more, everyone was still relying on basically the same technology—graphic processing units, or GPUs—that had been introduced back in 2010 by Laszlo Hanecz, the buyer of the Bitcoin pizzas. By the end of 2012 there was the equivalent of about 11,000 GPUs working away on the network.

But even back in 2010 it had been clear that if Bitcoin became more popular there was a logical next step that would eclipse GPUs. An application-specific integrated unit, or ASIC, is a chip that is built to specifically accomplish just one task—an even more specialized computing unit than a GPU. If someone could build an ASIC designed specifically to solve the Bitcoin hash function, it would probably be able to crunch the numbers hundreds of time faster than a GPU and thus likely to win hundreds of times more Bitcoins.

But designing and fabricating a new ASIC chip could cost millions of dollars, and take several months, requiring contracts with one of the five specialized chip foundries that produced virtually all the chips in the world. For most of 2011 and 2012 Bitcoins simply were not worth enough to justify this investment.

But as Bitcoin's price had continued to rise in the second half of 2012, a couple of enterprising engineers had thrown caution to the wind and begun racing to create the first ASIC chip dedicated to mining Bitcoins. The first entrant in the race was a company in Kansas City that went by the name Butterfly Labs. In June 2012 the founders announced that they would deliver specialized mining computers installed with custom chips in October 2012 and quickly sold $5 million of the machines on preorder.

A few months later, when Butterfly announced that the release of its machines would be delayed, a young Chinese immigrant in New York, Yifu Guo, announced that he had created a company, Avalon, with a group of engineers in China, which was building its own Bitcoin-dedicated ASIC chips.

Yifu, a shaggy-looking twenty-three-year-old, promised that each device would be able to do 66 billion hashes per second, compared with the 2 billion that a GPU card could do. What's more, his chips required a lot less energy—and thus lower electricity costs—to do the work. The price for each machine? A cool $1,299.

The process of putting the machines together, first in Beijing and then in Shanghai, and then shipping them to customers in the United States, proved to be more complicated than Yifu and his team anticipated. But on January 30, 2013, Jeff Garzik, the Bitcoin developer in North Carolina, posted on the forum pictures of the bulging boxes that DHL had just delivered and the gleaming silver box inside, built to do nothing but mint new Bitcoins. Within hours, new Bitcoins were showing up in Jeff's wallet, and within nine days the machine had earned back what Jeff had paid for it. The machine was eating up so much energy that it was heating up the room that it occupied.

Over the next month and a half, as the rest of Avalon's first batch of three hundred mining computers reached customers, the effect was evident on the charts that tracked the power of the entire Bitcoin network. It had taken all of 2012 for the power on the network to double, but that power doubled again in just one month after Yifu's machines were shipped. At the same time, the network automatically adjusted the difficulty of the problem the miners needed to solve, to ensure the ten-minute gap between new blocks of Bitcoins. For people who had built up fleets of GPUs making a profit quickly became a lot harder.[*]

A few other companies were making big promises about their own, specialized mining chips that they were working on. But the most aggressive project—and the one that revealed the most about the untapped potential that many saw in Bitcoin mining—was top secret and open to only a small elite. The company 21e6— shorthand for 21 million, the number of Bitcoins to be released— was created by Balaji Srinivasan, a Silicon Valley prodigy who had founded a successful genetics testing company from his Stanford dorm room. In the spring of 2013, Balaji was quietly assembling a team of top engineers to build a Bitcoin mining chip that would go beyond anything that had been contemplated before—rolled out in data centers built exclusively for the 21e6 machines. If the chips worked as promised they would mint money for investors. This was a simple enough proposition, and the price of Bitcoin was rising fast enough that it attracted interest from venture capitalists

---

[*] *See the Technical Appendix on page 357 for more detail on how the mining process worked.*

who were still queasy about tying their firms to Bitcoin. Both of the founders of Andreessen Horowitz, Marc Andreessen and Ben Horowitz, signed up to put some of their own personal money into Balaji's project, as did several of the original founders of PayPal, including Peter Thiel and David Sacks. Soon enough, Balaji was closing in on a $5 million fund-raising round.

The Bitcoin arms race had begun.

THE TYPE OF chip was not the only thing about Bitcoin mining that had changed since late 2010. Over the course of 2011 and 2012, more and more users were joining collectives that pooled their mining power. These mining pools allowed lots of people to combine their resources, with each person getting a proportional fraction of the total winnings, thus increasing the chances that everyone would get something every day.

The pools, though, generated concern about the creeping centralization of control in the network. It took the agreement of 5 percent of the computing power on the network to make changes to the blockchain and the Bitcoin protocol, making it hard for one person to dictate what happened. But with mining pools, the person running the pool generally had voting power for the entire pool—all the other computers were just worker bees. As a couple of pools harnessed significant computing power, some people worried that the operators of those pools could conspire to change or undermine the rules of Bitcoin.

But an incident in March 2013—the network's most significant technological failure to date—was a reminder of how the incentives built into the Bitcoin network could still work as Satoshi had hoped. Gavin Andresen, now the chief scientist of the Bitcoin Foundation, was in his den in Massachusetts after dinner, when he saw some online chatter about disagreement between computers or nodes on the network over what block the nodes were trying to mine—was it the 225,430th block since the network began back in 2009, or the 225,431st?

Gavin quickly realized that this was what had long been known as the biggest potential danger to the Bitcoin network: a "hard fork," a term coined to describe a situation where one group of computers on the network went off in one direction, agreeing about which node had mined each block, while another group of computers on the network moved in another direction, agreeing on a different set of winners for each block. This was disastrous because it meant that there was disagreement about who owned which Bitcoins. So far, there had been a split only on the last few blocks—not the whole blockchain history—but if it wasn't fixed, there would essentially be two conflicting Bitcoin networks, which would be likely to result in no one trusting either of them, or Bitcoin itself.

"this seems bad," a user on the chat channel wrote a few minutes after the problem first appeared.

" 'seems' is putting it lightly," another shot back.

"We have a full fork," one of the most respected developers, a Belgian programmer named Pieter Wuille, pronounced a few beats later.

The price of Bitcoin dropped from $49 back to $45 in a half hour, erasing all the previous week's gains.

Mark Karpeles joined the discussion a half hour later, and quickly stopped processing all transactions at Mt. Gox; a few minutes after that, Erik Voorhees said his gambling company, SatoshiDice, was doing the same.

By the time Gavin entered the conversation, it was clear that the problem was not the result of one node overpowering the network or of any sort of malice. Instead, computers that had downloaded a recent update to the Bitcoin software were accepting blocks—and awarding new Bitcoins to miners—that were not considered legitimate by the old software and the computers still running it.

Generally, if a block was accepted by a majority of nodes, it would be accepted by everyone, but the old software, version 0.7, had a rule that specifically did not allow a type of block that the new software, version 0.8, did allow.

The solution to this was clear: everyone on the network had to agree to move en masse to one of the two versions and adopt the blockchain accepted by that software. But there were no rules for deciding which version to pick—and once a version was chosen, no one knew how long it would take for all the nodes to get on board.

After racing through the possibilities, Gavin concluded that the most fundamental rule of Bitcoin was the democratic principle that the blockchain with the most support was the official one. In this case, the version created by the new software, 0.8, had a lot more computing power behind it. That was, in no small part, because the most sophisticated miners, especially the large pool operators, had been among the first to update their software. Gavin thought that if they had the most power, everyone else needed to update to join them. In addition to having more power, the miners on the new software had newly generated coins that they would be unlikely to want to give up.

Gavin quickly faced resistance from almost everyone else involved in the conversation; most participants believed that only the large miners would be responsive enough to change their software to fix the problem. Somewhat surprisingly, the operators of the biggest mining pools quickly agreed that they would revert to the old software, version 0.7. The operator of the prominent pool BTC Guild said that just switching his pool alone would get a majority of the computing power back on the earlier software. Doing this would mean losing the Bitcoins that had been mined since version 0.8 came out. But the losses would be much greater if the entire Bitcoin network lost the confidence of users.

"There is no way the 0.8 chain can continue in this situation," the operator of BTC Guild, who went by the screen name Eleuthria, said.

The developers on the chat channel thanked him, recognizing that he was sacrificing for the greater good. When he finally had everything moved about an hour later, Eleuthria took stock of his own costs.

"It could've been worse if I hadn't been able to start moving back to 0.7 quickly." But, he wrote, "this fork cost me 150–200 BTC"—over $5,000.

For the broader Bitcoin ecosystem, the price had fallen to $37, some 20 percent, within a few hours, and some online reports struck an ominous note.

"This is a dark day for Bitcoin. Implications for the exchange rate will likely be huge," a site called The Bitcoin Trader announced.

The incident had indeed revealed the sort of unanticipated problems that frequently occur in decentralized networks, which rely on lots of different members, with all their vagaries, acting independently.

But almost as soon as Eleuthria had fully switched his servers over to version 0.7 the price began recovering, and within hours people were talking about how the event had actually demonstrated some of Bitcoin's greatest strengths. The network had not had to rely on some central authority to wake up to the problem and come up with a solution. Everyone online had been able to respond in real time, as was supposed to happen with open source software, and the users had settled on a response after a debate that tapped the knowledge of all of them—even when it meant going against the recommendation of the lead developer, Gavin. Meanwhile, the incentives that Satoshi Nakamoto had built into the network had again worked as intended, encouraging people to look out for the common good over short-term personal gain.

•

A WEEK LATER, Gavin was back at his desk in the den not long after
dinner, when an unexpected announcement popped up. It came
from the Financial Crimes Enforcement Network, or FinCen, the
division of the Treasury Department responsible for monitoring
money laundering and enforcing the Bank Secrecy Act. In opaque
bureaucratic terms, the release stated its intent to "clarify the applicability of the regulations implementing the Bank Secrecy Act
('BSA') to persons creating, obtaining, distributing, exchanging,
accepting, or transmitting virtual currencies."

Reading behind the legalese, Gavin could see that this was
the United States government's first statement on the legality of
Bitcoin.

"oh wow," Gavin Andresen wrote on the chat channel before
passing along a link to the announcement for everyone else.

Everyone had feared that at some point the authorities would
step in and declare virtual currencies illegal. As Gavin and others
furiously scanned the lengthy document, the doomsayers were
quick to give their read.

"this kills the Bitcoin," one user on IRC responded to Gavin.

But as Gavin and others read on, they saw that it was not, in
fact, all bad. Yes, the document noted that anyone selling virtual
currency for "real currency or its equivalent" would now be considered a money transmitter—a category of business subject to lots
of stringent federal rules. But the release also made clear that many
parts of the virtual-currency universe—including miners—were
not subject to these regulations. More important, Jeff Garzik, the
programmer in North Carolina, noted, the basic implication of the
message cleared up the biggest single cloud: "this solidifies Bitcoin
status as legal to possess and use for normal people."

Indeed, Gavin said: "More legal/regulatory certainty is definitely a good thing . . . even if we might not like the regulations."

Over the next few days, Bitcoin companies all raced to understand the specifics of the FinCen guidance. Exchanges clearly needed to register as money transmitters, but what about companies like BitInstant that just worked with exchanges? And did exchanges also need to register as money transmitters with each state, as companies like Western Union had to do?

In New York, Charlie got an e-mail from one of BitInstant's lawyers: "I don't think this is good for the community."

But for the broader Bitcoin universe, the basic message of the guidance was encouraging: the United States government was not planning to come in and shut down the virtual currency. The next day the price of Bitcoin surged from $52 to $59, and by Thursday it was above $70.

The financial crisis sweeping Europe added yet another boost to the price. The banks on the Mediterranean island of Cyprus were on the verge of collapsing in mid-March when European authorities put together a bailout plan. The hitch was that all savings in Cyprus's banks were to be docked by 10 percent. The government, in other words, was confiscating money from private bank accounts. *BusinessWeek* ran a story that conveyed the seeming promise of Bitcoin: "BITCOIN MAY BE THE GLOBAL ECONOMY'S LAST SAFE HAVEN," the magazine's headline said. Russians who kept their money in Cyprus's banks were rumored to be buying up Bitcoin, which no government could confiscate.

The prices certainly suggested that someone with lots of money was buying. In California, Wences Casares knew that no small part of the new demand was coming from the millionaires whom he had gotten excited about Bitcoin earlier in the month and who were now getting their accounts opened and buying significant quantities of the virtual currency. They helped push the price to over $90 in the last week of March. At that price, the value of all existing coins, what was referred to as the market capitalization, was nearing $1 billion.

On March 27 the forums and the news site Reddit lit up with calculations of what value, for a single coin, would take the market capitalization over $1 billion, and the number settled on was $91.26. This calculation was largely theoretical: most of the outstanding coins had been purchased for pennies or a few dollars in the early years, and if everyone tried to sell for $91, the price would plummet. But it marked a psychological line in the sand that was, if nothing else, fun to talk about. That day, Cameron Winklevoss, who had taken responsibility for the twins' buying and selling of Bitcoins, was watching the price closely, first from the twins' office and then from home. After midnight, as he was preparing to go to bed, he saw the price approach the magical border of a billion. As the number crept closer and closer, he placed a small order on Mt. Gox that would be executed only if the seller agreed to a price above $91.26. The order was quickly filled and he watched the value of a Bitcoin on Mt. Gox—determined by the last order—jump to $91.27. Twitter and Reddit went wild. The next morning, Cameron gleefully reported to Tyler that it was their money that was responsible for sending the value of all Bitcoin over $1 billion for the first time.

CHAPTER 20

●

March 2013

The surging price of Bitcoin helped bring out of the woodwork some of the early Bitcoiners who had dropped from view.

In February, Martti Malmi posted an entry on his company's website describing his early days in Bitcoin. A month later, Hal Finney recounted his own story on the Bitcoin forum. By this time, his ALS had progressed to the point where he was essentially paralyzed, relying on tube feeding and a respirator. He spent most of his time in the same living room where he'd first worked on Bitcoin four years earlier, his old computers stacked up on the desks around him. But Hal could still communicate and type using a computer that tracked his eye movement, and he diligently worked on a few coding projects and regularly checked in on Bitcoin to see how his pet project was doing. As he watched the price go up, he asked his son to burn the private keys to his Bitcoin wallets onto a DVD, and put the DVD in a safe-deposit box at a bank. Some of his coins, though, he had his son sell, in order to pay for all the medical care he needed to stay at home.

"I'm pretty lucky overall," Hal wrote. "Even with the ALS, my life is very satisfying. But my life expectancy is limited. Those discussions about inheriting your Bitcoins are of more than academic interest. My Bitcoins are stored in our safe deposit box, and my son and daughter are tech savvy. I think they're safe enough. I'm comfortable with my legacy."

THIS SHOULD HAVE been the best of times for the existing Bitcoin businesses, and in certain ways it was. In March alone, sixty thousand new accounts were opened on Mt. Gox, and the monthly trading commissions rose above $1 million for the first time ever, more than triple what they had been a month earlier.

But even after all their earlier struggles, the staffers at Mt. Gox were not ready for this surge in business. Mark Karpeles now had a staff of eighteen, and a deputy with real business experience, whom he put in charge of all the company's dealings with the outside world. But Mark gave this deputy no power over the company's actual operations and kept firm control of Mt. Gox's essential accounts. Mark also continued to struggle with prioritizing his responsibilities. He was two years into running the world's largest Bitcoin exchange, but he had still not attended a single Bitcoin event abroad—a fact that he blamed on the sickness of his cat, Tibanne, who needed daily shots that Mark believed only he could administer.

Meanwhile, in late 2012 Mark had agreed to hand over his American customers to Peter Vessenes and his company CoinLab, which had an American bank account. But when it came time to hand over the customer files in March, Mark flinched, worried about some of the terms in the contract he had already signed. This left Mark's customers relying on Mt. Gox's Japanese bank, which put strict limits on the number of wires the company could send out each day. Even the simple task of opening an account with Mt. Gox required a three-week wait for approval from Mark's team.

For BitInstant and other companies that had to work with Mt. Gox, the reason behind the problems seemed simple: sheer incompetence. Charlie Shrem's BitInstant was now the main driver of trading volume to Mt. Gox, but when there were problems Charlie's e-mails to Mark Karpeles would go unanswered for days or even weeks.

Wences Casares had never fully trusted Mt. Gox and had been looking for a better place to store his coins. When he put them into his own digital wallet, he realized that all his private keys—the signature that allowed his coins to be spent—were sitting on his computer or phone, waiting for the first hacker who got access to his computer. Someone who had the private key for one of Wences's Bitcoin addresses could, essentially, impersonate Wences. Wences decided to work on a system with his Argentinian friend Fede Murrone to store their private keys out of the reach of hackers. They started by putting all their private keys on a laptop, with no connection to the Internet, thus cutting off access for potential hackers. After David Marcus, Pete Briger, and Micky Malka put their private keys on the same offline laptop, the men paid for a safe-deposit box in a bank to store the computer more securely. In case the computer gave out, they also put a USB drive with all the private keys in the safe-deposit box.

CHARLIE HAD KEPT BitInstant ahead of the regulatory curve. Back in 2012 he had registered the company with FinCen as a money transmitter. In March, though, the company was still trying to bounce back from the departure of Erik Voorhees and his friend Ira Miller, who had moved to Panama to develop their own company after the falling-out with the Winklevoss twins.

The new team Charlie brought on immediately spotted significant flaws in the way the company was being run. For starters,
Charlie was the only one with access to the company's bank accounts. Many day-to-day operations required Charlie to manually
intervene. The new lead developer called for the entire site to be
taken down and rebuilt. But there wasn't time as new customers
were pouring money into the site. The new staff members were
jammed into every corner of the small offices Charlie and Erik had
moved into the previous summer.

On top of the internal problems, Charlie was also having trouble finding a reliable bank account, even as a registered money
transmitter. Since the end of 2012, Charlie had opened accounts
with KeyBank, PNC, Wells Fargo, and JPMorgan Chase—and
all of them had been shut down. It became apparent to others
in the company that Charlie had not been entirely up-front with
the banks about the nature of his business. Charlie had generally
opened the accounts without explaining that BitInstant customers would be depositing and withdrawing money on a daily basis.
When the banks saw the thousands of transactions every day—a
strain on their compliance officers—they decided the BitInstant
business wasn't worth it.

This pointed to a broader issue with Charlie that was frustrating
the Winklevoss twins and was clearly an outgrowth of his childhood
desire for acceptance. Charlie loved telling people what they wanted
to hear. He would always give the twins optimistic predictions for
projects and would fail to alert them to impending problems until
the last moment, in the hope that the problems would go away. This
optimistic approach was great for a salesman, and Charlie had been
a great salesman. But it was not such a great habit for a manager,
who needed to find a way to deal with problems, not ignore them.

Given the issues at Mt. Gox and BitInstant—the two longtime
giants of the Bitcoin world—investors and entrepreneurs in Silicon

Valley were looking for alternatives. As an alternative to Mt. Gox, people saw some promise in Bitstamp, an exchange that had been founded by a Slovenian college student and a family friend back in 2011 and that had been growing slowly ever since. Wences and Micky sent one of Micky's deputies to Slovenia to scope out the operations. The youngsters running Bitstamp looked like an Eastern European boy band, with their long hair and penchant for Adidas track suits. But their evident competence—particularly when they were compared with Mt. Gox—generated so much confidence that Wences and Micky began moving their trading to Bitstamp. Mt. Gox still had 80 percent of all Bitcoin trading, but Bitstamp's market share began to creep up.

For those looking to buy smaller quantities of Bitcoin—BitInstant's specialty—people found their way to Coinbase, a San Francisco–based startup that had been opened by a veteran of Airbnb and a former trader at Goldman Sachs at the beginning of 2013. The company had managed to interest several investors and had maintained a bank account with Silicon Valley Bank. But even with Coinbase executives at the bank made it clear that the Bitcoin business was testing their patience. In order to stay on top of anti–money laundering laws, the bank had to review every single transaction, and these reviews cost the bank more money than Coinbase was bringing in. The bank imposed more restrictions on Coinbase than on other customers because Bitcoin inherently made it easier to launder money. A terrorist could potentially put dollars into Coinbase, buy Bitcoins, and then use the blockchain to send those Bitcoins to terrorist cells overseas. Because there is no identifying information attached to Bitcoin addresses, the terrorist cell could receive money without anyone noticing. That is very different from a traditional bank, in which every account is tied to a specific person or organization. Coinbase had to repeatedly convince Silicon Valley Bank that it knew where the Bitcoins leaving

Coinbase were going. Even with all these steps, on several days in March Coinbase hit up against transaction limits set by Silicon Valley Bank and had to shut down until the next day.

At the end of the month, an item was posted on SVBitcoin, an invite-only e-mail list for the Silicon Valley Bitcoin community: "The Time Has Come for the Bitcoin Community to Own a U.S. Based Federally Chartered Bank."

The author, an investor named David Johnston, wrote that the skepticism of traditional banks toward virtual currencies was the biggest roadblock facing Bitcoin's growth. If people couldn't send dollars from their bank to BitInstant or Coinbase, the surging interest in virtual currencies would be snuffed out.

The community was hitting a roadblock that almost every movement striving to disrupt the status quo eventually reaches. The big ideals of Bitcoin had carried it a long way and were sound in theory, but eventually the community required some cooperation from the existing authorities—people needed the old banks to agree to move their money into the Bitcoin realm. This was like an anarchist commune that ran up against the unwillingness of local officials to continue delivering water and electricity. Such collisions with the recalcitrant real world are frequently where utopian schemes run into trouble.

Johnston estimated that purchasing a licensed bank that could specialize in Bitcoin companies would require something like a $10 million investment up front. He offered to put up $1 million himself—thanks to the big rise in his own Bitcoin holdings—and he sought out ten more investors to join him. Charlie quickly wrote back saying it was a great idea. Wences responded next, offering to help fund the venture.

But there wasn't time for any big changes. On April 1, 2013, the price of a Bitcoin crossed the $100 threshold, a 670 percent increase since the beginning of the year.

The price moves were now feeding on themselves, as speculators chased the climbing ticker, fueled by news articles from all the new acolytes, many of them tutored by Wences. Jeremy Liew, a venture capitalist at the firm Lightspeed Capital, which had money in Wences's current startup, wrote an article in TechCrunch explaining that: "As a VC, my interest in the *Bitcoin* ecosystem is not ideological but mercenary. I see the opportunity for Bitcoin to disrupt multi-billion-dollar markets, but in doing so also create new big markets."

Within the companies handling all the money, however, the gaskets popping and wood warping were once again audible. Charlie didn't have enough money at Mt. Gox to fill all the orders coming in. On April 5, with the price moving above $140, he asked the Winklevosses for a short-term loan of $500,000 so that he could increase his reserves.

"I really wanna make 4pm wire cutoff so I can make sure we have enough money for the weekend in our accounts!" Charlie wrote feverishly.

When they quickly sent over the funds, he wrote back: "Thanks guys, you are amazing."

Charlie was also running into issues at Mt. Gox, where he purchased many of the coins that he sold to BitInstant customers. With orders pouring in, Mt. Gox was so backed up it was taking half an hour for trades to go through. This exacerbated the price swings as people who thought they were buying at $160 weren't getting their coins until the price was at $175.

To compound matters, Mark Karpeles chose this moment to move ahead with big changes in some obscure but important codes that customers used to transfer money around, and did not fully brief customers—even big ones like BitInstant—on how to cope with the changes. This set off a set of increasingly panicked e-mails between Tokyo and New York.

"You have been throwing us around like you always do," Charlie wrote to Mark on April 9. "Beating around the bush and not being up front with us."

When Mark responded without answering Charlie's basic question about some necessary coding language, Charlie exploded: "IF WE CANNOT ACCEPT MTGOX ORDERS WE ARE VIRTUALLY SHUTTING DOWN."

"Someone help us!!!!" Charlie wrote on the morning of April 10.

That same day, the mania peaked when the price for Bitcoins on Mt. Gox surged to $260. In the first ten days of the month, the exchange had attracted 75,000 new accounts. On April 10 the number of trades coming in was three times higher than it had been just a day earlier. For a trade, the lag between being entered and being executed was more than an hour. As people sat waiting for their orders to go through, they saw the price shoot up and panicked, not wanting to pay $300 when they had intended to pay $200. Orders were canceled and people began to sell, hoping to lock in the profits they had realized over the past months. The effect was predictable. While Charlie was asleep in New York the price began crashing, and by the time Charlie showed up at the office, the price was down to $200. By lunchtime it was closer to $100.

The BitInstant engineers congregated with their laptops on the small black sofa and chairs in Charlie's office. Charlie had a bottle of rum on his desk and, in a spirit of good fun, was taking occasional swigs as everyone tried to figure out just what was going wrong. Even the wireless network in the office was failing because of the number of people in the building trying to help. When Yifu Guo, the creator of the Avalon mining chips, stopped by the office, Charlie was in a state of giddy panic, both scared and amused.

"I'm flipping out. I'm yelling at everyone. Yifu, I'm drinking the rum from the bottle," he said with a laugh.

"I don't know why you guys are all freaking out," Yifu said, chuckling himself. "I'm not worried. The price is fine. It's time to buy."

Things calmed down for a few hours after Mark Karpeles assured his users that the problems were due to the volume of trade, not to hackers. But hours after he wrote that, the hackers showed up and staged fierce denial-of-service attacks, forcing Mark to shut down the site altogether in the middle of the day.

CHAPTER 21

●

April 11, 2013

The day after Mt. Gox shut down under the strain of heavy trading, the members of the corporate board of Lemon, Wences Casares's digital wallet, showed up at the company's Palo Alto offices for a lunchtime meeting. The price of Bitcoin sat more than 50 percent lower than where it had been twenty-four hours earlier. But the sudden downturn had done nothing to dim Wences's faith in Bitcoin's future. Instead, it had increased his conviction that the companies dominating the Bitcoin universe, like Mt. Gox, needed to be replaced, and that he needed to do more than just be a cheerleader for Bitcoin among the Silicon Valley elite.

Lemon provided a way for customers to keep all their credit cards and coupons in digital form on their smartphones. Wences proposed to his board that they add a pocket for Bitcoins that would be a safe, reliable way to keep virtual currencies and potentially even to buy them. To get started, Wences suggested that Lemon could use $1 million of its remaining money to buy Bitcoins that could serve as an initial pool for customer purchases. This was actually a great time to buy coins, Wences argued, because the price was down after the latest price crash.

Wences expected to see his board members light up—particularly Micky Malka, the chairman of Lemon's board and one of the first people Wences had gotten excited about Bitcoin back in 2012. Instead, Micky furrowed his brow. Is this really what Lemon set out to do? Micky asked Wences. Lemon had finally started catching on as a digital wallet. Wouldn't opening it up to virtual currencies engender all sorts of unknown legal risks?

The other board members quietly listened to Wences's explanation of why this was worth doing. They all knew it was dangerous in Silicon Valley to alienate an entrepreneur like Wences—there was no easier way to ensure that a company failed. But they didn't jump to his defense either.

After the meeting was adjourned, the board member who had looked the least skeptical, Eric O'Brien, pulled Wences aside and asked him: "How strongly do you believe in this—what are you personally doing?"

Wences didn't mince words: "I am personally allocating a percentage of my net worth to this that is borderline irresponsible because I believe in it so much."

Regardless of what the Lemon board wanted to do, Wences said, "I would advise you to invest as much money as you can stomach losing."

He told O'Brien to buy coins at Mt. Gox, but to move the coins off Mt. Gox as soon as the order went through.

"It is either going to be worth zero or worth five thousand times what it is today."

IN THE DAYS that followed, Mt. Gox reopened for business and the price stabilized around $100. But many believed that the recent price crash proved the flaws in the whole concept. Felix Salmon, a financial columnist at Reuters, wrote a widely circulated article pointing out that the volatile price of Bitcoin made it nearly impossible to use for its most basic purpose, as currency. If consumers didn't know whether a Bitcoin would be worth $10 or $100 tomorrow they would be unlikely to spend their coins and merchants would similarly be unlikely to accept them. Even this critic, though, saw something elegant in the network underlying Bitcoin.

"For the time being, Bitcoin is in many ways the best and cleanest payments mechanism the world has ever seen," Salmon wrote. "So if we're ever going to create something better, we're going to have to learn from what Bitcoin does right—as well as what it does wrong."

The day after the crash, the Winklevoss twins finally went public in the *New York Times* with their now significant stake in Bitcoin—worth some $10 million. The interest was not restricted to the United States. A few weeks after the crash, a national television station in China broadcast a half-hour segment on the new enthusiasts in that country, and several local entrepreneurs began setting up exchanges to buy Bitcoins using yuan.

Despite the crash, everyone with a Bitcoin idea found that there was now no shortage of eager investors in Silicon Valley. In May, Pete Thiel's Founders Fund announced that it was putting $2 million into BitPay, the payment processing company that allowed merchants to accept Bitcoin and end up with dollars in their bank—taking advantage of the Bitcoin network's quick and cheap transactions.

But the company that was attracting the most attention was Coinbase, founded by the veterans of Airbnb and Goldman Sachs. The twentysomething cofounders had clean-cut looks and soft-spoken ways that naturally engendered confidence. Investors liked that the pair avoided the ideological talk of overthrowing the Fed and instead sold their company as a safe and easy place for consumers to buy and hold coins that wouldn't be subject to endless delays and scrutiny from the authorities. They also had real professional experience at well-known companies, something that had been in short supply in the Bitcoin world up to this point.

After consultations with Wences, Micky decided to team up with the New York venture capitalist Fred Wilson to put $5 million into Coinbase. It was the largest publicized investment in a Bitcoin company to date, by a wide margin, and the first time an established venture capitalist like Wilson had put serious money into the space. The rest of Silicon Valley took notice.

CHARLIE, MEANTIME, WAS taking advantage of BitInstant's status as the only serious Bitcoin company in New York—the media capital of the world—to become a sort of public spokesman for Bitcoin in the press. He regularly invited reporters to a bar that he had invested in at the beginning of the year, EVR, the sort of dark, swanky Manhattan club that made its clientele line up in high heels on the sidewalk. The round leather booth in the back corner was Charlie's standing nighttime office, with some top-shelf liquor on the table for guests.

Those who knew Charlie back in Brooklyn were amazed at his transformation from a short, awkward teenager into a confident impresario who bragged about the ring that he wore, engraved with the private key to one of his Bitcoin wallets. But as always with Charlie, it was all somewhat less than it appeared. He still lived in his teenage bedroom in the basement of his parents' house in Brooklyn. He left people with the impression that EVR was his bar, despite the fact that he had put in only about $15,000 and owned less than 1 percent of it. Meanwhile, Charlie's company was racing furiously to keep up with all the new competitors, especially Coinbase, and Charlie was often missing when he was needed most, hanging out at the bar or talking with reporters. At one point, Cameron

Winklevoss asked Charlie: "Do you want to be the proprietor of a bar or the CEO of a Bitcoin company? You can't have it both ways."

Cameron, the more involved of the two twins, constantly pressed Charlie on why things weren't getting done faster. When Coinbase's $5 million investment was announced, Cameron warned Charlie that Coinbase could steal BitInstant's thunder.

"Just deliver the deliverables and stop fucking around," Cameron told him.

Charlie meekly submitted. "OK, I will push the team and myself harder."

In Tokyo, Mark Karpeles was also learning that Mt. Gox's first-mover advantage was not impregnable.

On May 2, Mark was sued in a Seattle court by CoinLab, the company run by Peter Vessenes that had been scheduled to assume responsibility for Mt. Gox's business in the United States earlier in the year. CoinLab accused Mt. Gox of breaching its contract by not handing over the customers. Troubles deepened a week later when the money in Mt. Gox's two American bank accounts—some $5 million—was seized by federal agents, who accused Mt. Gox of violating federal money-transmitting laws. It wasn't apparent at the time, but these moves were part of the net tightening around Silk Road, as law enforcement agents in Baltimore narrowed in on their prey. Prosecutors had secretly filed a sealed indictment on May 1 against Dread Pirate Roberts for narcotics trafficking and were prepared to arrest the mastermind as soon as they figured out who he was.

Given all this turbulence, it was remarkable that anyone continued using Mt. Gox at all. In the world of trading, though, the most valuable thing an exchange can offer is liquidity or, more simply, people buying and selling. An exchange with the best technology in the world isn't worth anything if no customers are there offering to buy and sell. Mt. Gox still had liquidity because it had attracted so many customers from its days as just about the only exchange around, and some customers would move only if others did as well.

But a chasm was opening up between the early Bitcoiners and the new, more practical community of entrepreneurs, engineers, and investors. When some of the developers working on the underlying Bitcoin code set up a Bitcoin press center, it immediately led to fights about who was presentable enough to be listed as a contact for journalists, especially when Roger Ver was taken off the list. Erik Voorhees lashed out at those trying to smooth Bitcoin's sharper edges.

"It is embarrassing to see Bitcoin reduced to sniveling permission-seekers, too cowardly to speak about the real issues and the real reasons why this technology is so important," Erik wrote. "Bitcoin is a movement, and those trying to distill it into nothing more than a cute new technology are kidding themselves and doing a terrible disservice to this community."

EVERYONE SEEMED READY for a truce from the bickering as the Bitcoin Foundation's first-ever conference approached in late May. The foundation had booked the main convention center in the capital of Silicon Valley, San Jose.

On the morning of the conference's first day, Friday, May 17, the Valley news site *TechCrunch* went live with a story that officially announced the investment the Winklevoss twins had made in BitInstant, which had remained a secret even after they went public with their holdings of Bitcoin. The investment was put at $1.5 million. Even this article was the cause for a small tiff with Charlie, who had accidentally tipped off another reporter first.

"Your communication is piss poor and gums up the entire operation," Tyler Winklevoss wrote.

But the tension quickly passed and Charlie and the twins showed up at the convention center to find that they were heroes of sorts to the assembling Bitcoin masses. Many of the conference attendees had been aficionados for years, waiting for the world to see the beauty of their pet project. Now these tall, statuesque celebrity twins were standing up for their cause. On Friday night, the twins delivered the keynote speech together, clad in sneakers and button-down shirts with rolled-up sleeves. They opened with a quote from Gandhi, and proceeded to cite Dr. Seuss and the Bitcoin pizzas purchased by Laszlo Hanecz. The next morning, when the general exhibition opened, one vendor was selling shirts with the smiling face of Charlie Shrem, in the style of Barack Obama's famous "Hope" poster.

The adulation distracted Charlie from the business opportunities at the conference. He got around to scribbling down some thoughts for his Saturday afternoon speech only an hour beforehand, while standing around the booth. The talk was unsurprisingly disjointed, but Charlie still possessed his old infectious enthusiasm, which had the crowd cheering and clapping. That night, the whole BitInstant team went out for a boozy dinner with shots of Fireball whiskey, followed by a night out at a club.

While Charlie and other Bitcoin old-timers were reveling, a more quiet and sophisticated conversation was going on around the edges. In a back room of the convention center, Gavin Andresen gathered with the four other developers who were maintaining the basic Bitcoin software that computers on the network were running. This was the first time the so-called core developers had met in person, and far from the crowds they talked about the serious work of keeping the basic Bitcoin protocol safe from hackers and forks.

The moneyed set that had recently converted to Bitcoin was also buzzing around the conference. Wences didn't speak at the conference but he had lots of private conversations with the investors and entrepreneurs whom he had introduced to the technology, including PayPal's David Marcus, who had turned his name badge around so that no one would know who he was. After browsing in the exhibition hall, Marcus told Wences that he had been appalled by the naïveté and lack of sophistication of the existing companies. When asked how they were dealing with anti–money laundering laws, none of the young entrepreneurs gave a knowledgeable answer. It was so bad that Marcus told Wences he was contemplating quitting PayPal and starting his own Bitcoin exchange—something he later decided against.

For these Silicon Valley power brokers, there was an absurdity to the old-school Bitcoiners who crowed to each other about being the leaders of a new global movement and getting rich in the process. The convention center happened to be hosting the Big Wow! ComicFest at the same time as the Bitcoin conference, and it was sometimes hard to tell who among the long-haired nerds were there for the comics and who for the virtual currency.

●

CHAPTER 22

June 2013

The gap that had been revealed at the Bitcoin Foundation's conference—between the apparent promise of Bitcoin's underlying idea and the weakness of the current companies—only emboldened the big-money people going into the summer.

Pete Briger at Fortress, the private equity and hedge fund giant, invited in an old classmate from Princeton and colleague from his days at Goldman Sachs, Dan Morehead, to help Fortress look full-time at a range of virtual-currency opportunities. A tall, statuesque man, who had been on both the rowing and the football teams at Princeton, Dan looked like a member of the ruling class, and he had recently been running his own hedge fund, Pantera. After getting the invitation from Briger, Dan took up a desk at Fortress's offices in a skyscraper near the Embarcadero in downtown San Francisco. He soon hired the first professional traders to buy Bitcoins for a fund he hoped to set up, which would make Bitcoin more easily available to big investors. In New York, Barry Silbert was working on something similar. To get everyone in his company involved and excited, Barry gave each of his seventy-five

employees two Bitcoins—each worth around $100 at the time—
with the mandate to spend one of them and save the other.

But as these professionals got more deeply involved it quickly
became clear to them that for all the excitement around Bitcoin
in Silicon Valley, almost no one had been paying attention to
equally important constituencies in Washington, DC, and on Wall
Street, now the most significant roadblocks to the growth of this
technology.

In late May federal prosecutors arrested the operators of Liberty Reserve, another online currency that Mt. Gox and BitInstant
had used early on as a method for funding accounts. Liberty Reserve was a very different beast from Bitcoin. It was run by a centralized company, which designed the currency to make it easier
for criminals to move money undetected. But the shadow of Liberty Reserve naturally fell on Bitcoin and statements from regulators suggested they did not necessarily see a big difference.

At the end of May the top financial regulator in California sent
the Bitcoin Foundation a cease-and-desist letter accusing the foundation of operating as an unlicensed money transmitter. The accusation was somewhat absurd—the foundation was not a business
of any sort—but it highlighted just how little the foundation had
done to cultivate relationships with the relevant regulators.

Given the regulatory uncertainty, it was unsurprising that
bankers were not eager to get involved with the new industry. In
2012 and 2013 several big banks had faced $1 billion fines for
not being vigilant enough in tracking money laundering. In the
early summer of 2013 JPMorgan Chase, the nation's biggest bank,
was shutting down accounts for any companies that came with an
elevated risk of money laundering, including check-cashing businesses and companies that did remittance payments to Mexico.

Finding banks willing to open accounts for Bitcoin companies
had always been a problem for entrepreneurs like Charlie Shrem.

But even the new, more powerful backers of Bitcoin were discovering that they couldn't find banks willing to work with them. Fortress's Pete Briger set up a meeting with top executives he knew at one of the nation's largest banks, Wells Fargo, about potentially teaming up to create a more secure and reliable Bitcoin exchange, but Wells Fargo quickly declined any partnership. It had been only a few months since Wells Fargo had had to deal with federal agents seizing Mt. Gox's Wells Fargo bank accounts.

In all the discouraging dealings with bankers and government officials, Bitcoiners were facing basic questions about why it was worthwhile for anyone to put any energy into this technology. Almost five years after Satoshi Nakamoto had published his paper, the virtual currency was worth real money and had attracted talented people, but although some small companies accepted Bitcoin through BitPay, the virtual currency was still used almost entirely for speculation, gambling, and drug dealing.

Economists who had taken note of Bitcoin also pointed out that the virtual currency actually had built-in incentives discouraging people from using it. The cap on the number of Bitcoins that could ever be created—21 million—meant that the currency was expected to become more valuable over time. This situation, which is known as deflation, encouraged people to hold on to their Bitcoins rather than spend them.

The notion of Bitcoiners around the world sitting on their private keys and waiting to become rich begged the question of the intrinsic value of these digital files. What were all these locked-up virtual coins really worth if no one was doing anything with them? What was backing up all the value the coins seemed to have on paper?

Bitcoin fans argued that the United States dollar was not backed up by anything real either—dollars were just pieces of paper. But this argument ignored the fact that the United States government promised to always take dollars for tax bills, which was a real value no matter how much people disliked paying taxes.

Practically no one was promising to take Bitcoin for anything. The primary value the coins had at this point was the expectation that they would be worth more in the future, allowing current holders to cash out for more than they paid. To some cynics, that description made Bitcoin sound suspiciously like a less savory sort of financial invention: a Ponzi scheme.

FROM THE OUTSIDE, it would have been easy to conclude that Charlie and BitInstant were somehow dodging all these problems. Charlie was shopping for new, larger real estate for his company and eventually settled on a well-appointed suite in an office tower. Charlie had finally managed to move out of his parents' basement in Brooklyn. He was motivated to do this, in no small part, because he was afraid to tell his parents about his girlfriend, Courtney, who was a waitress at his favorite bar, EVR. Courtney was some ten years his senior and, more important, not Jewish—something that did not fly in the Syrian Jewish community. Charlie and Courtney took a room in a big communal apartment above EVR, where there were always alcohol bottles and bongs on offer. Charlie was often spotted at EVR with Courtney on his arm.

But within BitInstant, Charlie's hard-partying ways seemed to many like an escape from the challenges he was facing with his company. The Winklevoss twins had been pushing Charlie to raise more money to pay for BitInstant's expansion. And Charlie had no trouble getting meetings with investors, who were all impressed at the sheer number of dollars already running through BitInstant. But as Charlie's team tried to get the investors the paperwork they needed, it quickly became clear how unequipped BitInstant was for the big time. When the BitInstant chief financial officer, who was just two years out of college, tried to put together the financial statements he realized that there were large holes in the company's books, with unexplained expenses in all directions.

Charlie had made remarkable progress for a twenty-three-year-old entrepreneur with almost no prior experience. He had built a complicated business from nothing and people entrusted him with millions of dollars. But Charlie was clearly, and unsurprisingly, lacking skills as a manager. In many startups this is something that investors might notice, and help fix, by finding an experienced manager to come in and steer the ship. As it turned out, though, Charlie's investors didn't have much more experience working with startups than he did. The twins' early experience with Mark Zuckerberg had been limited and, since setting out to become tech investors the previous year, they had worked with only a few young companies. With Charlie, the twins had initially adopted a hands-off attitude, despite all the bickering. But as problems became more evident, they talked with Charlie's chief programmer about replacing Charlie as CEO. When Charlie learned about the potential palace coup he was furious and began showing up for work less and less.

In mid-June, the Winklevosses asked an angel investor they knew, Chris Morton, to diagnose BitInstant's problems. What they got back was a long list of basic things the company was missing, among them:

"There is no accounting system.

"The equity agreements are a mess or nonexistent.

"The company mission is not clear."

But Morton's harshest words were reserved for Charlie:

He cannot focus. He seems to be busy with superfluous meetings (press, investors, partners, speaking engagements) and personal commitments (bar, rental property). Even when those meetings are in progress, he does other things

on his computer. He makes commitments and does not
follow through. He confirmed a meeting with the accoun-
tant and then did not show.

The Winklevoss twins talked with Morton about coming in to
help turn around the company, but he had little interest.

The twins were realizing that BitInstant might be a lost cause
and they began working toward a life in Bitcoin without Charlie. At
the Manhattan offices of Winklevoss Capital, where the brothers
had matching glass-walled offices on either side of a glass-walled
conference room, the twins started putting together the paperwork for what they envisioned as the first-ever Bitcoin exchange-
traded fund, or ETF, which would hold Bitcoins and move with
the value of the coins, but trade on a real stock exchange, much
like the hugely popular gold ETF. The twins planned to assemble a
team that would buy and sell Bitcoins, allowing ordinary investors
to purchase the ETF through their Charles Schwab or E*Trade
brokerage account.

IN LATE JUNE, Charlie finally managed a long-planned relaunch
of BitInstant, in partnership with a money-transmitting business
that was regulated in most states. But when the site went live and
BitInstant began doing more thorough checks of its customers,
Charlie's staffers realized that many of their customers had been
doing business with them under fake identities. When the Manhattan district attorney sent a disconcerting request to Charlie asking
him to come in for a meeting, it precipitated an emergency conference call with a team of lawyers on July 4.

"The problem is that the site is a patchwork of bandages," one
of the lawyers told Charlie and his team. "When we go into that

meeting, they're going to go straight to the site and review it in detail. They can't see a patchwork of quick fixes."

The lawyers were unrelenting, and the answers from Charlie made them nervous: no, BitInstant's compliance officer had no previous experience in compliance, and no, BitInstant had not filed any suspicious-activity reports with regulators despite having lots of transactions flagged as potentially fraudulent by partners. The call concluded with a long list of things that needed to be handled immediately.

"You are very exposed on all fronts," the lawyer told Charlie and his team.

Charlie tried to show how serious he was about complying with all the rules, but the old problems were quickly joined by new ones. A couple of customers disputing transactions filed a lawsuit, for which they were seeking class-action status. When the twins read Charlie the riot act, he responded with total contrition.

"Things ARE changing dramatically to fix problems on all fronts and put us in a position for growth as quickly as possible," he told them. "I've made a lot of mistakes, the ones that you guys called me out on as well as others that I'm seeing now and taking steps to fix."

But there wouldn't be time for that. Charlie was in the new BitInstant offices, which he had moved the company into less than two weeks earlier, when he got a letter from his lawyers telling him that because of the number of legal questions, they could not represent him in his upcoming meeting with the district attorney unless he shut down the site and resolved all the problems.

Charlie reached the Winklevoss twins while they were in the car on the way to their family beach house. They laid the blame entirely at his feet and demanded the return of the $500,000 loan they had made back in April when business was booming.

On Friday, July 12, at 9 p.m., Charlie took the BitInstant site down, for what he thought would be only a temporary hiatus.

THE MALODOROUS HAZE now hovering over Bitcoin was making everyone question what it was doing.

Erik Voorhees, one of the most fearless proponents of Bitcoin's radical possibilities, announced a few days after Charlie shut down BitInstant that he was selling the gambling site, SatoshiDice, which he'd bought in 2012 and turned into one of the most popular Bitcoin sites on the Web.

The sale involved reimbursing all the people who had bought shares in Erik's company in 2012, but they had only 13 percent of the site. This young man who had been unemployed two years earlier was now a millionaire living in Panama. But the reason he was selling SatoshiDice was not the money. In e-mail exchanges with other entrepreneurs he explained that his legal costs were piling up and that it was too much of a headache to be under such scrutiny.

"Bitcoin businesses are literally at the edge of law, not because they are doing anything wrong, but because Bitcoin enables new activities and behaviors and recategorizes money in such a way as to enable it to transcend current statutes. This is both exciting, and scary, because we're breaking amazing ground and we'll inevitably be in the crosshairs for doing so," he said.

About a week after he sold the company and paid back his shareholders, he got an e-mail from the Securities and Exchange Commission letting him know that it believed that he had broken the law by selling unregistered securities. The e-mail caused a terrible feeling in the pit of Erik's stomach that didn't abate for days.

Not long after that, nearly every major company in the Bitcoin space got a subpoena from the top financial regulator in New

York, a young bulldog of a prosecutor named Benjamin Lawsky, who asked for a trove of documentation about consumer protections and anti–money laundering programs. A few days later the US Senate's Committee on Homeland Security and Governmental Affairs sent a letter to the major financial regulators and law enforcement agencies asking about the "threats and risks related to virtual currency." Neither of these requests suggested that lawmakers regarded this new technology with much warmth.

No one, though, was feeling more heat than Ross Ulbricht, aka Dread Pirate Roberts.

Ross's site was more successful than ever. In the middle of 2013, Silk Road was approaching its one-millionth registered account. In the first two months of the summer, Silk Road users exchanged over a million messages with each other and the commissions collected by the site were often over $10,000 a day.

But since the spring Ross had been dealing with continuing and varied attacks unlike anything he had experienced before. A hacker had managed to take the site down for days at a time and stopped only after Ross agreed to pay $100,000 up front and $50,000 every week thereafter—payments that ultimately amounted to $350,000.

These weren't the only unanticipated costs. When a user named FriendlyChemist threatened to release details about thousands of Silk Road customers, Ross reached out to a distributor, who he believed was a member of Hell's Angels, and asked what it would cost to do away with FriendlyChemist. This time around, there was none of the hemming and hawing that had accompanied Curtis Green's supposed death. When the assassin, redandwhite, came back with a price of $150,000, Ross politely haggled with him.

"Don't want to be a pain here, but the price seems high," Ross wrote, pointing to the $80,000 that had been paid for the previous execution.

A few days after a price was agreed upon, redandwhite sent evidence that the deed had been done (though no evidence was later found of an actual murder). Messages quickly followed with a request for a hit on another scammer—and three of his associates—who had robbed Silk Road users. This deed was paid for with 3,000 Bitcoins, or roughly $500,000 (but, again, no evidence was found of any actual murders).

This was not the softhearted young man of early 2012 who had trouble telling white lies. Now his diary was filled not with ruminations on his weaknesses, but instead with brief, cold lists of his problems and solutions. His entry for the day FriendlyChemist was presumably killed, read:

> got word that blackmailer was excuted
> created file upload script
> started to fix problem with bond refunds over 3 months old

Even his family members, who had no idea what he was up to, noticed a change during this time. Ross's mom would say that her son, during this period, was "rebel Ross," not the lovable young man she had known in recent years.

Ross's transition from an affable youngster obsessed with oneness to a minor tycoon whose diary entries reflected a willingness to kill looked, from many angles, like a predictable outcome of the community that Ross had created and the role that Ross had assumed within that community. In a world in which there are no agreed-upon authorities, it was natural that individuals might take it upon themselves to determine what is right and wrong—and to act on those determinations on their own. It was easy to imagine that Ross, cut off from any real contact with the other members of the community, except for Internet chats, began to see people as abstractions with no real life force—like characters in a video game. In this sort of world, the idea of killing these people could lose its visceral repugnance.

As the year went on, Ross receded further from his ordinary life. He moved out of his friend's apartment in June and went even deeper underground, renting a place a few miles away in a residential neighborhood of San Francisco that he paid for in cash. He told his new roommates that his name was Josh. On his laptop, he kept a document called "emergency" that included the steps he would take if he needed to run:

> encrypt and backup important files on laptop to memory stick.
> destroy laptop hard drive and hide/dispose
> destroy phone and hide/dispose
> hide memory stick
> get new laptop
> go to end of train
> find place to live on craigslist for cash create new identity (name, backstory)

The New York office of the FBI was, by this point, working in cooperation with the Marco Polo task force that had been set up a year and a half earlier in Baltimore to crack down on Silk Road. The teams were making almost monthly arrests of other vendors and buyers on Silk Road, and many of these arrests were publicized. When a competing black market drug site, which had opened in the spring, shut down, Dread Pirate Roberts told his followers that he had often thought about doing the same:

Without going into details, the stress of being DPR is
sometimes overwhelming. What keeps me going is the understanding that what we are doing here is more important
than my insignificant little life. I believe what we are doing
will have rippling effects for generations to come and could
be part of a monumental shift in how human beings orga-
nize and relate to one another.

I have gone through the mental exercise of spending a
lifetime in prison and of dying for this cause. I have let the
fear pass through me and with clarity commit myself fully
to the mission and values outlined in the Silk Road charter.

Ross, by this point, understood just how hard it was going to be
to continue evading detection. He became aware, at several points
in 2013, that despite his best efforts, his system did occasionally
leak a real IP address, providing information, however briefly, on
where his servers were located. Each time, he would delete the
information and move his databases to new servers, hoping that
no one had noticed the mistake. Ross assigned Variety Jones, his
old mentor, who now went by the screen name cimon, to serve as
the site's counterintelligence expert against law enforcement. But
as Ross guessed, there were, indeed, federal agents dedicating their
days to spotting any sign of a real IP address associated with Silk
Road, and they were homing in on a set of servers in Iceland that
they believed were the right ones.

Before the authorities got anything on those servers, though,
agents on the Canadian border intercepted a package with nine
forged drivers' licenses. Each license had a different name and address, but the pictures on all of them were the same wavy-haired
young man. The package was addressed to a house in San Francisco. When agents knocked on the door, they recognized the young
man from the photos on the forged IDs. He quickly presented his

real driver's license, from Texas, with his real name, Ross Ulbricht. He declined to answer any other questions about where the IDs had come from, but told the agents in an offhand way that anyone could buy faked documents from a site called Silk Road.

The agents left without taking Ross with them. He had gotten lucky. While he was one of the suspects that the New York and Baltimore agents were looking at, they had not disseminated his name widely, and the border patrol officers had no idea who he was. After this close call, Ross changed apartments, but he did not take the opportunity to cut and run. Instead, he stayed in San Francisco, watching his commissions from Silk Road pour in as the digital noose tightened around his neck.

PART THREE

●

CHAPTER 23

August 2013

The Bitcoin Foundation had set out to help improve the network's public standing, but most of the people involved in the foundation's creation had now become unhappy examples of the technology's problems. Charlie Shrem had shut down his site and was being sued. Peter Vessenes was locked in a legal battle with his fellow founding board member, Mark Karpeles, and Peter's other ventures were going just as poorly. A company he had set up to produce Bitcoin mining machines had not yet turned out a single coin and his investors were breathing down his neck.

There was, though, one unlikely person left to carry on the original mission of providing the technology with a more friendly public face: the Seattle lawyer Patrick Murck. For most of 2012 and 2013 Patrick had worked for existing Bitcoin companies and volunteered as general counsel of the foundation. But since the beginning of the summer he had been employed by the foundation full-time and was turning himself into a respectable public spokesman.

At each point along the way, Bitcoin's survival had required the strengths of a different subset of its believers. In the summer of 2013 it had become clear that if Bitcoin was going to reach a larger audience it would need to learn how to play nice with the existing system. As it turned out, Patrick, a pudgy young father with a warm fuzzy beard, was uniquely positioned to do just that. In contrast to Bitcoin's early salesmen, like Roger Ver, who was still trying to renounce his citizenship, Patrick was a patriot who had grown up in Washington, DC, with a mother who worked at the National Labor Relations Board. This upbringing had made him believe in the importance of fighting injustice in the world and gave him a healthy respect for the role that government could play in the process, which helped explain the volunteer work he had done for the Obama campaign in 2008.

When it came to Bitcoin, Patrick firmly believed, like many in the tech world, that Bitcoin could foment big changes. An open source financial network looked to Patrick like just what was needed to shake up the privileged elite who ran and disproportionately benefited from the existing financial system. The Bitcoin network seemed to make it at least a little bit harder for Wall Street to collect tolls at every step of every financial transaction. But Patrick did not think that for this to happen it would be necessary for Bitcoin to overthrow the existing governments and central banks. In fact, he thought there was a significant place for regulations when a third party, like Mt. Gox or BitInstant, was holding someone's virtual currency.

Patrick had quietly begun his work at the beginning of the summer, when he spoke at a conference in Washington that represented essentially the first time a Bitcoiner had sat on the same stage with lawmakers. At that point, there had been obvious tension. Patrick had ended up in a sharp exchange with a man from the Department of Justice who had compared Bitcoin users to child pornographers.

Afterward, though, Patrick struck up a friendly conversation with the woman in charge of FinCen, the branch of the Treasury Department that had put out the first rules on virtual currencies in March 2013. Patrick had been somewhat peeved that FinCen and its leader, Jennifer Shasky Calvery, had not had any conversation with the Bitcoin community before issuing those rules. At the June conference, though, Shasky Calvery made it clear that she was interested in the technology and open to a dialogue about the rules.

Over the course of the summer Patrick made almost weekly trips from Seattle to Washington to meet with Shasky Calvery and other regulators, to help them understand Bitcoin. Patrick quickly learned that staffers in the office of Senator Thomas Carper, of Delaware, were studying Bitcoin and looking at the possibility of holding a hearing. Patrick was able to put them in touch with the most presentable players in the Bitcoin world.

In his meetings Patrick did not fight the obvious reality that Bitcoin was not yet doing any of the great things that he and others were talking about. But he was able to cogently explain his vision of how the blockchain technology could make it easier for poor immigrants to transfer money back home and allow people with no access to a bank account or credit card to take part in the Internet economy.

In addition to his legal mind, Patrick had a genial, unthreatening approach that made him able to get along with just about anyone. He liked having his conversations over a whiskey or beer in a bar, and his everyman sensibility tended to soften people up. The good relationship Patrick developed with Shasky Calvery, among other people, led to a private meeting in August, when Patrick and a few other people affiliated with the Bitcoin Foundation got to present Bitcoin's best face to a roomful of law enforcement agents and government officials. It was not entirely friendly, but the attendees seemed to understand that the Bitcoin technology was useful for more than just purchasing drugs and laundering money—so this meeting was already a long way from Patrick's first encounters in Washington at the beginning of the summer.

Many Bitcoin companies were making their own efforts to get in sync with the authorities. Coinbase, the San Francisco–based company that had raised $5 million from Micky Malka's Ribbit Capital and other investors, was developing extensive measures to vet clients and ensure that the service was not used toward illegal ends. The Slovenian Bitcoin exchange, Bitstamp, which passed Mt. Gox over the summer to become the largest Bitcoin exchange in the world, now required all its customers to go through a rigorous identity verification process. The two young men who had founded the exchange were rewarded with visits to their Slovenian city, Kranj, by Dan Morehead and Pete Briger from Fortress Capital, who wanted to invest in the exchange.

THIS WAS ALL a long way from the original Cypherpunk vision of a new digital money that was outside the reach of governments and banks. Satoshi Nakamoto's aim in creating the decentralized Bitcoin ledger—the blockchain—was to allow users to control their own money so that no third party, not even the government, would be able to access or monitor it. But people were still opting for the convenience of centralized services like Coinbase and Bitstamp to hold their coins.

The great benefit of this business model was that the companies, rather than their customers, dealt with the headache of storing and securing the money. When early Bitcoin users lost the private keys to their Bitcoin addresses, the coins associated with those addresses were lost forever. With a Coinbase wallet, on the other hand, if a customer lost the password, it was like losing the password to a normal website—the company could recover it. What's more, Coinbase customers didn't have to download the somewhat complicated Bitcoin software and the whole block-chain, with its history of all Bitcoin transactions. This helped turn Coinbase into the go-to company for Americans looking to acquire Bitcoins and helped expand the audience for the technology.

There was, though, a small but vocal community of dissidents, many of them early Bitcoin users, who were eager to go back to the original vision that Satoshi had laid out. Few were as outspoken as Roger Ver, the Tokyo-based libertarian who had, in earlier years, lost money that he had entrusted to Bitcoin businesses like Bitcoinica and MyBitcoin.

Roger was still a fervent believer in the initial vision he had of Bitcoin as a game-changing technology for governments around the world, just as his favorite martial art, jujitsu, offered a relatively simple way to neutralize even the strongest opponent. Roger had recently begun comparing Bitcoin to the honey badger, the weasel-like equatorial mammal that has a reputation for being able to overpower and even castrate the most ferocious predator. During the summer of 2013, with graphic design assistance from Erik Voorhees, Roger had put up a new billboard in Silicon Valley with a picture of the indomitable animal and the caption: "Bitcoin: The honey badger of money."

But Roger had grown increasingly firm in his belief that centralized Bitcoin businesses like Coinbase defeated the essential purpose of Bitcoin by putting the personal information of every user in the files of a single company that was vulnerable to government subpoenas. In the summer of 2013, aiming to foster an alternative, Roger channeled the energy that he had earlier put into Charlie Shrem and BitInstant into another one of the startups he had invested in back in 2012.

Blockchain.info had been created by a reclusive young man named Ben Reeves who lived in the English city of York and ran his site alone until the middle of 2013. Reeves had created what looked like a rather unspectacular product: an online wallet that, like other wallets, offered a way to access Bitcoins from any computer or smartphone without downloading the entire blockchain. But Reeves's wallet was different in a crucial way. Rather than holding its customers' Bitcoins, Blockchain.info kept only a small file for each customer with the private keys of that customer, encrypted in a way that made it impossible for the company to see the keys themselves. Because Blockchain.info held an encrypted file with the keys, they were not on the computer of the user, vulnerable to hackers. But when a customer logged into a Blockchain. info wallet, the log-in process decrypted the file so that the keys were temporarily on the customer's computer and could be used to access coins that the customer had on the blockchain. The customer's data—how much money he or she had and the transaction history—was viewable through Blockchain.info's online template. But the company itself never saw the data. Because Blockchain. info did not hold money or a transaction history for its customers, it couldn't be subpoenaed to give up customer records. Nor could the company steal its customers' coins.

The site had attracted lots of interest from people who opened 350,000 free Blockchain.info wallets by the middle of 2013. But the business model was not a recipe for big profits. Because Blockchain.info didn't hold customer funds it was hard to deduct fees for its services. It also didn't allow its customers to buy Bitcoin online—a lucrative business that would have put the company in charge of customers' money. Blockchain.info users had to acquire their coins elsewhere and send them to their Blockchain.info wallet.

This was a business opportunity uniquely suited to Roger Ver, who had never been concerned, primarily, with making money from his Bitcoin investments. He wanted to see Bitcoin live up to its revolutionary potential. As a result, when Reeves offered to turn a loan that Roger had made to Blockchain.info into a majority stake in the company (so that Reeves could avoid a tax headache), Roger jumped at the opportunity.

In London for a conference that summer, Roger paid for Reeves to come down so they could meet in person for the first time. Reeves showed up, but Roger had trouble getting more than a few words out of the shy young man. After Roger went out to speak at the conference, he came back to his hotel room and found that Reeves had abruptly left and gone home to York.

This didn't discourage Roger. He thought Reeves's code spoke for itself and he began looking for a chief executive for the company, a person who could deal with the outside world so that Reeves didn't have to. When Erik Voorhees put Roger in touch with an old college fraternity brother, Nic Cary, Roger flew Nic to Tokyo. On their first night, they went to Roger's favorite establishment, the Robot Restaurant, where women in blinking bikinis rode around on large robotic animals. Roger and Nic spent the next few days immersed in deep conversations—some of them during drives around Tokyo in Roger's Lamborghini— about how to expand Blockchain.info's offering of a wallet that could be used free by anyone, anywhere in the world, outside the reach of regulators. Nic explained his vision for making the website more user friendly and expanding the number of languages.

Roger promptly hired Nic to move to York and work with Reeves in a three-story house that Roger rented for what he hoped would soon be a much larger team. As Roger began to build out the company he determined that this would be a real Bitcoin company, with no bank accounts and all salaries paid in Bitcoins.

TO MANY REGULATORS and investors, the only plausible reason that someone would want an untraceable Bitcoin wallet, like Blockchain.info, was to enable online drug purchases or other nefarious activity. Why else would you want to keep your records from government officials?

But one place where Blockchain.info, and Bitcoin more broadly, was gaining popularity in the summer of 2013 put a slightly different slant on the potential uses for Bitcoin services that couldn't easily be monitored by the government.

At a Bitcoin Meetup in July 2013, two hundred or so people packed into one of the historic old buildings that fill downtown Buenos Aires, the capital of Argentina. At a time when Bitcoin's popularity was faltering in the United States, the turnout in Argentina was many times greater than the thirty or so people who had attended the most recent meetups in New York and Silicon Valley. Many of the attendees in Buenos Aires had come looking for an easy way to buy Bitcoins and those who purchased coins from other attendees, generally with cash, were usually set up with a Blockchain.info wallet to receive their coins.

This was a long way from the first Bitcoin meetup in Argentina, which had been organized by Wences Casares back in 2012 and had attracted only a handful of people. Since then, Wences had given the credentials for the meetup group to one of his old friends who lived in Buenos Aires. Each of the meetups that his friend, Diego, organized had attracted more people, with a big jump in July. The increasing interest was not hard to understand in the Argentinian context. Over the first half of 2013, the Argentinian peso had been plummeting in value against other currencies.

While the government tried to deny the rampant inflation, grocery prices surged and everyone tried to dump pesos. The government's increasingly desperate attempts to keep money in the country—by imposing a tax on foreign credit card transactions, for instance—only made the problem worse. Keeping savings in pesos was equivalent to throwing the money away, but the government made it hard to get money out of the peso through official channels. This made a currency like Bitcoin and a wallet like Blockchain.info, which the government could not access, very attractive.

In late June, one of the nation's largest newspapers, *La Nación*, had put a story about *dinero digital* at the top of the front page of the Sunday issue. The people quoted in the article made it clear why they were interested.

"You don't have to be battling all of the government's problems, you aren't going to buy bread with it, but it'll save you if you have a stash of stable currency that tends to appreciate in value," twenty-two-year-old Emmanuel Ortiz told the newspaper.

Bitcoin, with its famous volatility, did fall in value against the peso in May and June 2013, when the problems at Mt. Gox created widespread pessimism. But by the end of the summer, Bitcoin had risen in value against the peso every other month of the year, and in September it was up 860 percent against the dollar since the beginning of the year while the peso was down some 25 percent against the dollar.

The excitement was building in Argentina despite the fact that the government's strict control over the financial system made it all but impossible for Argentinians to buy coins from an online service like Coinbase or Bitstamp. But Argentinians were used to figuring out less-than-official ways to deal with the government's twisted financial policies. The most prominent signs of this, during normal times, were the black market money changers— known as *arbolitos*—who were a regular presence in downtown

Buenos Aires. For Bitcoin, a similarly informal network of money changing was developing. A few of Wences's friends, including Diego, offered to meet up with people in person to exchange pesos for Bitcoins, turning themselves into the first digital money changers. The vision that Wences had back in 2012—of an online gold that offered Argentinians an alternative to the peso—was beginning to come true.

WHILE PEOPLE CLOSE to Wences were leading the charge in Argentina, Wences himself did not have time to think much about his homeland. He was too busy dealing with the problem that he faced with his digital wallet, Lemon.

Since the spring, Wences had been trying to find ways to integrate Bitcoin into Lemon and had been looking for investors to support him. The people excited about Bitcoin asked why they should put their money into a company like Lemon, which Wences had been struggling to get off the ground for two years. Perhaps more dispiritingly, Wences was unable to bring around the existing board of Lemon, and particularly his chairman and old friend, Micky Malka.

"These people didn't invest in a Bitcoin company," Micky would tell Wences about the Lemon investors. "What they invested in you created and it has value, and you are deciding for them to do something they would prefer not to do, which is throw it in the trash and do a Bitcoin company. If you want to do it, they will follow you, but that wouldn't be their preference."

Micky's continued resistance over the course of the summer left Wences with an unfamiliar sense of uncertainty. He did not want to give up Lemon—he had put too much energy into it and felt he owed it to his employees and investors to see it through.

What's more, he had long ago told his wife that he would not do another startup. But Lemon was not his true passion, Bitcoin was, and he felt he was missing out every day he was not working on it full-time. Wences's chiseled face carried lines of discontent that his friends had not seen before.

In September he went to a number of his closest friends to ask for their advice. One of those friends, a banker at Allan & Co., expressed surprise that Wences hadn't reached this point sooner.

"You are too successful and too wealthy to do things that aren't your passion," this friend told Wences.

When Wences told his friend about the obligation he felt he had to Lemon's employees and investors, the friend frowned in disagreement and told Wences that if Lemon could be sold it would allow the employees to continue working on Lemon while also getting money back to investors.

"You aren't an indentured servant to these people," the friend said. "If you can land the plane, it's good for the employees and you can reboot with something new."

After hearing something similar from another trusted friend, Wences went to his wife, Belle, and asked her what she thought. Belle surprised Wences by fully siding with his friends.

"You need to stop everything you are doing and do Bitcoin," she told him.

"But Belle," he said, "it's going to be another startup."

She wasn't listening to it: "I've never seen you so intensely held by something."

Wences immediately began offering Lemon around. He found that lots of big-name companies, including Facebook, PayPal, and Apple, were interested in buying Lemon, but only if Wences stayed on board. Wences turned them down. He didn't need the money they were offering him—the Bitcoins he had bought when they cost a few dollars each were now worth tens of millions of dollars, in addition to his previous wealth. More important, he was now certain that his primary goal was to be able to work on Bitcoin full-time. Another company that was pursuing Wences, the security company Lifelock, offered to buy Lemon and let Wences go pursue his passion. He quickly began the paperwork to get his board's approval and free himself.

CHAPTER 24

●

September 30, 2013

T he spinning top that had been Ross Ulbricht's life for much
     of the last three years was wobbling out of control in late
     September. He was trying to chase down the truth of a tip
he'd gotten about one of his most prolific vendors getting busted.
At the same time, Ross was angling to arrange a meeting with
redandwhite, the user who had been hired as an assassin earlier in
the year. Ross lent redandwhite $500,000 so he could become
a vendor on the site, but recently redandwhite had disappeared.
Meanwhile, when Silk Road's biggest imitator and competitor,
Atlantis, shut down, the operators of the site told Ross they'd
heard that the FBI had found a way to crack the anonymity of Tor.
To add insult to injury, while he was trying to get a piece of trash
out of a tree near his apartment in San Francisco, he got covered
in poison oak.

"I have poison oak rash from head to toe," he wrote to an old
girlfriend in mid-September. "I wish you were here to comfort me :("

On the last day of September, he wrote in his diary that he was
taking steps to get his life back under control: "Had revelation

about the need to eat well, get good sleep, and meditate so I can stay positive and productive."

It would be his last journal entry.

The next day he spent the morning working at home on his Samsung 700z laptop. In the early afternoon, he left home in his jeans and T-shirt, with his computer in a bag tossed over his shoulder. He made the quick five-minute walk, past the local BART transit station, to one of his favorite haunts with good wifi, Bello Café. When he walked in and saw how crowded it was, he turned around to go next door to the local branch of the San Francisco Public Library. He did not take any particular notice of the two men sitting on a small metal bench across the street, one of them holding a Mac laptop.

Ross walked across a narrow alleyway and upstairs to the newly renovated library, which sat above a gourmet grocery store. He headed to the far side of the library, away from the reference desk, where he chose a seat next to the science fiction section, looking out a window at the cute commercial strip across the street. He took his laptop out and went through the laborious process of logging into his carefully secured computer, onto the library's public wifi, and through to the encrypted programs he used to run Silk Road. When he opened the encrypted chat program, Pidgin, that he used to communicate with his staff he saw that one of his newer moderators, cirrus, had just pinged him: "Are you there?"

cirrus was the Silk Road member who used to go by the name scout. Early in the year, Ross had convinced scout to become a staff member by pointing out how unlikely it was that they would ever be caught.

"sure, someone could stand behind you w/o you realizing it," he had said back then. But he said the chances of that were "incredibly small."

On this Tuesday afternoon, cirrus asked Ross—or dread, as he appeared on cirrus's screen—how he was doing.

"dread: im ok, you?"

"cirrus: Good, can you check out one of the flagged messages for me?"

"dread: sure"

"dread: let me log in."

To get to the flagged messages, Ross signed into his administrative account on the Silk Road marketplace, an account that he had nicknamed mastermind. While he was getting in, he passed the time by asking about cirrus's past work exchanging Bitcoins. When cirrus told Ross that he had stopped the work because of the "reporting requirements," Ross shot back: "damn regulators, eh?"

Finally Ross was into his account, and the plain-looking boxes on the screen showed just how successful the business still was. There were 25,689 orders in transit from the site's 1,468 vendors. In his own administrative account, Ross had 50,577 Bitcoins, worth some $6.8 million at that day's exchange rate of around $140.

"ok, which post?" he asked cirrus.

This was the signal that cirrus had been waiting for. It told cirrus that Ross was now logged into the fortified inner sanctum of Silk Road. cirrus was, in reality, one of the men who had been sitting on the bench across from the café, Jared Der-Yeghiayan, a federal agent with the Department of Homeland Security. Der-Yeghiayan had convinced the woman who had previously been cirrus—and before that, scout—to hand over the account to the authorities.

Der-Yeghiayan was still outside, now with his computer open, and when he saw Ross's words pop up on his screen, asking him which flagged post cirrus was referring to, Der-Yeghiayan made sure to keep the chat with Ross alive, but he also signaled to the FBI agent sitting next to him, who in turn, signaled to a team inside the library.

Sitting at his computer, Ross heard a man and a woman fighting behind him. "I'm so sick of you," the woman shouted.

As Ross turned around to see what was happening, he saw out of the corner of his eye that someone swooped in on his table and grabbed his open laptop. Before he could turn around and do anything about it, several other people who had apparently been browsing in the stacks came at him and pinned him against the window. After he was handcuffed, other people who had been milling around the library converged on him and quickly walked him down the stairs and outside, where he was put into an unmarked van and read his Miranda rights. The plainclothes federal agents milling around outside the van had flown to San Francisco over the previous days. They came in from the many offices around the country that had been chasing Ross—or Dread Pirate Roberts—for months, and in some cases, years.

Ross didn't know it at the time, but his downfall had not come through the sophisticated hacking techniques and leaking IP addresses that he had worried about so much. The Internal Revenue Service agent who finally identified Ross did so by searching on Google through old posts on the Bitcoin forum. There the agent found a single job advertisement that Ross had placed in late 2011, under the screen name altoid—the account he had used to post the first ad about Silk Road in early 2011. The job ad from altoid was seeking someone who wanted to be a "lead developer in a venture backed Bitcoin startup company." The post had told interested applicants to contact "rossulbricht at gmail dot com." This was the one time Ross had connected his own e-mail address with altoid and Ross had realized his mistake and deleted it. But his e-mail was captured in the forum posting of someone else who had responded to Ross, leaving his name out there for the search engines. As much as Ross had wanted to create a new world, he still had to occasionally interact with the old one, searchable by Google, and that, rather than any mistakes in the new world, was what did him in.

The next morning, as Ross sat in a cell in Glenn Dyer Jail in Oakland, federal prosecutors in New York and Baltimore unsealed their own cases against him in federal court. The charges included narcotics conspiracy, conspiracy to commit computer hacking, and money laundering conspiracy, as well as an accusation that he had solicited a murder for hire to protect his site—the $80,000 he had allegedly paid to kill Curtis Green back in January. Almost any of the counts, individually, could lead to a life sentence.

"Silk Road has emerged as the most sophisticated and extensive criminal marketplace on the Internet today," the New York complaint said.

> The Government's investigation has revealed that, during its two-and-a-half years in operation, Silk Road has been used by several thousand drug dealers and other unlawful vendors to distribute hundreds of kilograms of illegal drugs and other illicit goods and services to well over a hundred thousand buyers, and to launder hundreds of millions of dollars deriving from these unlawful transactions.

Users of Silk Road visiting the hidden site that morning, hoping to score some heroin or pot, found an FBI emblem over the announcement: "THIS HIDDEN SITE HAS BEEN SEIZED."

WHEN ROSS'S ARREST was made public at around noon, New York time, on October 2, Cameron and Tyler Winklevoss were sitting together, with their laptops, at the dining room table in their family vacation home on Long Island.

It was an unseasonably warm day and they had spent the morning in the ocean on their paddleboards. They were no longer spending time on BitInstant, but they were still building up their stash of virtual currency and working with regulators to get approval for the Bitcoin ETF. At the dining room table where they had done their initial research on Bitcoin a year earlier, they read through the Silk Road indictment as they watched the price of Bitcoin begin to fall.

There had never been a reliable accounting of how much Silk Road was driving the overall Bitcoin market. But many of the headlines that the Winklevoss brothers read out to each other assumed that illegal transactions were a major force in Bitcoin that would now go missing.

"I just hope that mainstream adoption has surpassed the adoption of criminals and drug dealers. LOL! Otherwise its time to SELL! SELL! SELL!" one forum user wrote.

Selling is what a lot of people were doing, sending the price down to $110 from $140 within two hours after the news came out.

The panic was, of course, much worse on the Silk Road forums, where users were assuming that the government now had access to computers with information about every single Silk Road customer and vendor.

But the Winklevoss twins saw an opportunity. The best analysis they had seen suggested that Silk Road accounted for no more than 4 percent of all Bitcoin transactions, hardly a driving force. More important, they knew that Silk Road was one of the biggest black marks holding Bitcoin back with ordinary people, who assumed the blockchain was just a payment network for drug dealers. This arrest could help sever Bitcoin's association with crime.

The criminal complaint itself stated explicitly that prosecutors did not view the cryptocurrency only as a tool for breaking laws.

"Bitcoins are not illegal in and of themselves and have known legitimate uses," the FBI agent, who drew up the complaint, wrote.

This brief sentence was one of the strongest statements to date about the legality of Bitcoin in the United States—and it came from one of the divisions of the government most likely to want to shut Bitcoin down.

The twins didn't want to buy coins while the price was still dropping, but when they saw it begin to stabilize, Cameron, who had done most of the trading, began placing $100,000 orders on Bitstamp, the Slovenian Bitcoin exchange. Cameron compared the moment to a brief time warp that allowed them to go back and buy at a lower price. They had almost $1 million in cash sitting with Bitstamp for exactly this sort of situation, and Cameron now intended to use it all.

The twins were not the only people to seize this opportunity. About an hour after the price fell to $110, a surge of buying pushed it back above $130. By the time Ross was brought to court on Friday for a bail hearing, the price was just a few dollars shy of the $140 mark, where it had been before his arrest. In court, Ross was in shackles and wore a red prison jumpsuit. He said little and showed no obvious emotion. His publicly assigned lawyer said that Ross denied all the charges. The judge began preparations for moving Ross to New York, where he would await trial.

ON THE SAME day as Ross's court appearance in San Francisco, a very different side of Bitcoin was on display at a gathering south of the bay. Some of the most influential Bitcoin players were gathered at the San Carlos Airport outside San Jose. They were there

to board privately chartered flights to Truckee, California, the
closest town to Dan Morehead's vacation house on the shore of
Lake Tahoe.

Morehead had been helping Pete Briger examine the Bitcoin
opportunities available to Fortress. He had set up a sort of mini
hedge fund that would buy and hold Bitcoins and sell shares to
rich investors, while also looking to make investments in Bitcoin
startups. In October, he invited leading virtual-currency advocates
to his home in Tahoe for the first-ever Bitcoin Pacifica, a weekend
of socializing and conversation about his favorite digital money.

Among the people boarding the planes were the two founders
of Bitstamp. Morehead had paid to fly them in from Slovenia and
was hoping to finalize a $10 million investment in the exchange.

Roger Ver was in from Tokyo and spent most of the weekend
in a sweatshirt he had made with a picture of two honey badgers
copulating. Roger also brought along Nic Cary, the young man he
had hired to run Blockchain.info. Morehead was pushing Roger to
sell part of his stake in Blockchain.info, which was coming to look
increasingly valuable.

Morehead had also roped in Neal Stephenson, the author of
the science fiction book *Cryptonomicon*, which had popularized
the idea of virtual currencies when it was published in 1999. Roger
quickly got Stephenson set up with his first Bitcoin wallet, from
Blockchain.info.

Wences Casares couldn't make the trip to Lake Tahoe—he was
too busy closing the sale of Lemon—but his longtime collaborator,
Micky Malka, made the journey. Jesse Powell, Roger's old friend,
had volunteered to drive up to Morehead's house with a few people,
so that, in the event that Morehead's chartered plane crashed, there
would be a few people left to continue leading Bitcoin.

Once everyone was at Morehead's house, the conversations
predictably came back again and again to Silk Road. Few of the

attendees were pessimistic about what Ross's arrest would mean for Bitcoin. This seemed to many of them like the exact line in the sand that Bitcoin had needed to mark a division between its early, renegade years and its future in the mainstream. At dinner in Morehead's enormous living room, Roger sat with Briger and Nejc Kodric, the chief executive of Bitstamp. The men placed their bets on where the price would be in a year. While Briger was somewhat cautious, betting that the price would fall to $120, slightly below where it was that day in October 2013, Nejc guessed that it would be thirteen times as much, or $1,300, and Roger was even more optimistic, guessing $1,320.

CHAPTER 25

●

October 2013

The Cross Regions Plaza was an exemplar of the hastily built skyscrapers that littered the Shanghai landscape like so many gilded toothpicks. It had a lobby with gleaming marble floors and an entire wall covered by a leaping golden horse. But the elevator doors opened at each floor to reveal narrow, scuffed hallways reeking of smoke.

Just across from a smoking closet and next to the Yu Cheng Vacation Club, suite 23N was a small office, but still too big for the tiny staff it housed. Amid a few whirring upright black fans, one of the few people tapping away at a desk was a boyish, bespectacled thirty-year-old programmer, Huang Xiaoyu, who had recently moved to Shanghai from Hunan, where he had been living with his wife's family.

Xiaoyu had founded China's first Bitcoin exchange, BTC China, back in 2011 with the husband of his wife's college roommate, Yang Linke, who handled the nontechnical aspects of the company. It was Xiaoyu, on the Chinese-language Bitcoin forum, who had given Bitcoin its Chinese name, three characters that were pronounced *bee-te-bee*, a play off the Chinese word for currency.

Until recently Xiaoyu and Linke had run their exchange from opposite ends of the country as a sort of hobby, in time snatched from their real jobs. The small amounts of money moving into and out of the exchange went through the personal bank accounts of Linke. Nothing more was needed to sustain the light volume on the one and only exchange where Bitcoin could be bought and sold for yuan.

That had all changed owing to the commanding presence in suite 23N—a thirty-eight-year-old man with a stout, penguinlike body, and a wide face with round curious eyes. Bobby Lee, who generally wore the same khaki pants and blue dress shirt day in and day out, alternated between flawless English and imperfect Shanghainese as he explained his vision for Bitcoin's potential in the world's most populous nation.

WHEN BOBBY LEE had first reached out to the founders of BTC China in February 2013, he was much less well known in the Bitcoin world than his younger brother, Charlie Lee, the California-based Google engineer who had been involved in Bitcoin since 2011 and who was perhaps best known as the creator of Litecoin, one of the most successful alternative virtual currencies. It was Charlie who had pushed Bobby, and the rest of his family, to first look at Bitcoin back in 2011.

Bobby had a natural interest for the same reasons as his brother. The two men, who grew up sharing a bedroom, had both studied computer science, Charlie at MIT, Bobby at Stanford. Perhaps more important, both grew up in the Ivory Coast as the children of Chinese immigrants who had escaped the communist revolution with only the wealth they had stored in gold. When Bobby and Charlie roomed together in Silicon Valley, soon after college, Charlie had gotten Bobby into collecting gold coins and buying precious metals online. They understood cryptography as well as the importance of easily transferrable places to keep money.

Bobby, though, was less of a programming whiz than his little brother and had spent much of his career as a manager. His jobs in the e-commerce divisions of Yahoo and Wal-Mart had afforded him a comfortable life in Shanghai, where he and his wife lived in a gated, manicured community of apartment towers. But after years of working for someone else, Bobby had developed a hankering, common among many older brothers, to run something himself. And Bitcoin looked increasingly attractive in a Chinese context.

Bobby recognized that Chinese people would have little interest in the libertarian ideas of American Bitcoiners—decades of state-sponsored communism had killed most interest in ideologies. But after six years in Shanghai, Bobby believed that Bitcoin could have a unique, thus far untapped appeal in China. The most convincing evidence that it could take off was China's previous experience with a successful virtual currency, Q coin, a digital money launched in 2002 by a Chinese online company. Q coin had started as a way to buy digital goods like greeting cards, but by 2006 Chinese people were buying and selling the coins themselves, bidding the price up. The frenzy did not stop until 2009, when the government stepped in and said that Q coins could be used only for their original purpose. To Bobby, it seemed that the main things holding Bitcoin back from becoming the next Q coin were the lack of good information about Bitcoin in Chinese and the lack of reliable places to buy coins.

With this history in mind, in early 2013 Bobby had begun talking with his little brother about doing some sort of Bitcoin startup together. Charlie could do the coding and Bobby, as the more outgoing and confident brother, would be in charge. At the same time, to broaden his options, Bobby e-mailed the founders of BTC China. After using the exchange for many months, Bobby thought it had the potential to expand and improve. Within a few weeks of his first e-mail, plans were afoot to meet in Beijing, where the business-minded cofounder, Linke, lived. (Bobby had already become so excited about the prospect of working on Bitcoin that he turned down an offer to return to Yahoo.)

Xiaoyu flew to Beijing from Hunan and Bobby came up from Shanghai. During a dinner at Quanjude, a restaurant famous for its Peking duck, Bobby put it to Linke and Xiaoyu simply: if they would be willing to make him the cofounder and chief executive of BTC China, he would invest his own money and go out and raise funds to expand the company. He also said the company had to be based in Shanghai, given his wife's unwillingness to move from what she viewed as the most cosmopolitan city on the mainland. Bobby was not an easy person to say no to. He had a sincere demeanor that made it hard to doubt his honesty. His résumé also made it clear that he had about as many accolades as one could collect by the age of thirty-seven, including two degrees from Stanford and several years as an early employee at Yahoo.

Neither of the two cofounders of BTC China spoke English well or knew how to run a company, and both had been overwhelmed by even the small amount of business they had attracted. Bobby, meanwhile, had the perfect unthreatening teacherly way needed to introduce a foreign and potentially scary new concept. He explained things in careful steps, never speaking down to anyone. By April the founders had struck a deal for Bobby to join them.

A FEW WEEKS after Bobby signed his deal, Bitcoin had gotten its first major media exposure on the mainland, from China Central Television's Channel 2, which showcased just how immature the virtual-currency ecosystem was in China. The reporter for Channel 2 tracked down what he believed was the only place in the country that had accepted Bitcoins for purchase—an Internet café in Beijing, which had accepted its first Bitcoins at the urging of a young American expatriate living in the city.

But while there wasn't much visible activity in China of the sort that so many American entrepreneurs were pushing in the United States, there was quite a bit of work going on in the shadows. The reporter dug up a few young men who had set up fleets of computers with ASIC chips that were doing nothing but mining Bitcoins. Mining was a business that made sense in China, given the legions of tech-savvy youngsters and easy access to cheap electronics. But there was another, more systemic explanation for why the Chinese preferred less visible ways of acquiring their Bitcoins.

Like Argentina, China had incredibly restrictive rules about moving money into and out of the country. But in China, unlike Argentina, these rules were not a response to runaway inflation, but instead part of the government's effort to keep tight control over the exchange rate of the yuan, in order to promote the export economy. The authoritarian government also wanted to keep a close check on what its citizens were doing. Each Chinese citizen could move only the equivalent of $50,000 out of the country each year. As a result, it became difficult for wealthy people, including government officials, to get their riches out of China and into more secure foreign bank accounts.

Living in Shanghai, Bobby saw how capital controls did not just make it hard for rich people to hide their money in other countries. The controls also made it harder for China's rising middle class to invest in anything that wasn't Chinese. It was all but impossible to buy American or European stocks and bonds. This meant that ordinary Chinese investors eagerly latched onto every half-plausible new investment opportunity that presented itself.

Money had poured into the Chinese real estate and stock markets, pushing both into elevated territories that many thought were unsustainable.

Bitcoin presented an intriguing new investment that almost anyone with a computer could access. Bobby believed the Chinese would be all too willing to put their money into this unproved digital currency, despite the hazy legality—as the market for Q coins had demonstrated. Decades of communism had turned black markets into the norm.

There was also a more suspect explanation for all of this behavior and for Bobby's belief in his business. As a gambling man, Bobby knew that China was a nation of people with an unusual willingness to place a bet on just about anything. That is what made the Las Vegas of China, Macao, seven times bigger, in revenue terms, than Las Vegas. While Bitcoin's speculative nature and volatility were a strike against it in many countries, in China these had the potential to be its most attractive qualities.

OVER THE SUMMER, as the price of Bitcoin was stagnating, Bobby had raced to get his company set up for the next surge of interest. He went to the Bitcoin Foundation meeting in San Jose and looked for investors. In July he rented an office in Cross Regions Plaza, little more than a single room with two small conference rooms carved out with glass walls. The room looked down into the Shanghai National Stadium and out toward the hazy skyline sprawling into the distance.

Bobby's main focus was on striking a deal with the country's two major online payment processors—the Chinese counterparts of PayPal—so that BTC China's customers would have a way to get money into the exchange that didn't involve the personal bank account of the company's founder, Linke. The largest payment processor, Alipay, owned by the Chinese Internet giant Alibaba, was put off by the sound of Bitcoin, which it had not heard about before. But the smaller company, Tencent—not coincidentally, the creator of the old digital currency Q coin—was eager to provide something that Alipay didn't and signed up with Bobby in September.

In the United States, PayPal's unwillingness to work with Bitcoin exchanges had been a major hindrance. Once Bobby got Tencent integrated into BTC China's website in September, it was suddenly easier to get Bitcoins onto an exchange in China than it was anywhere else in the world.

Bobby was not the only one who had spotted the potential appeal of cryptocurrencies in China. During the summer of 2013, the number of people downloading the basic Bitcoin software in China had regularly been second only to the number in the United States, and mining operations continued to grow. By September two other exchanges were up and running with a full-time staff. But BTC China was already doing twice the volume of the other exchanges in the country, and Bobby Lee didn't intend to lose his early lead. He inked a deal to take a $5 million investment from Lightspeed Capital, the venture capital firm that had previously backed Wences Casares's company Lemon. Shortly thereafter, as a promotional tool, BTC China marked China National Day by removing the 0.3 percent commission that customers had to pay on every trade. In China, unlike anywhere else in the world, it was now essentially free to trade Bitcoin.

The real ascent in China began in mid-October, after the arrest of Ross Ulbricht, when a division of Baidu, the search engine giant and the fifth-most-visited website in the world, announced it would be accepting Bitcoin payments. A close look at the announcement revealed that it applied only to a tiny security service run by Baidu, Jiasule, but it gave Bitcoin a patina of legitimacy that it had so far lacked in China.

In the week after the Baidu announcement, the price of Bitcoin moved up not just in China but around the world, from about $140 to $200, with the volume of trading climbing faster in China than anywhere else. On October 19, forty thousand Bitcoins changed hands on BTC China, nearly twenty times the number that had been traded on most days in September. In mid-October, BTC China saw the most volume of any exchange in the world during a few days—the first time that any exchange other than Mt. Gox or Bitstamp had held this record. It was evident that China was leading the price up because the price was rising faster in yuan than it was in dollars. In Shanghai, Bobby began furiously hiring people to try to fill the space in his still half-empty office.

China was not the only source of momentum in the markets during this period. Many of the people who had attended Dan Morehead's gathering in Lake Tahoe had traveled on to Las Vegas for the Money 2020 conference, the same financial-industry conference that Roger Ver and Charlie Shrem had attended the year before. When Charlie was there, only one Bitcoin company had been exhibiting, BitInstant. This time around, Bitcoin companies flooded the exhibition hall and there were three different panels dedicated to the subject.

Then on November 3, the chief executive of eBay, John Donahoe, said in an interview with the *Financial Times* that PayPal was looking at creating a digital wallet that could eventually hold Bitcoins. After Donahoe's comments came out, the price, which had been hovering around $215, began rising, and three days later it surpassed the previous record price of $267 that had been set on Mt. Gox during the April pandemonium.

That same day, Bobby Lee was with his staff on a retreat to Shengsi Island. Much of the trip was spent trying to deal with the onslaught of new accounts and customer service requests. The pressure didn't relent for the trip back the next day. Bobby's exchange handled sixty thousand coins in one day for the first time ever, as the price leaped above $300 on Mt. Gox.

During this period, BTC China was seeing more trading volume than any other exchange in the world almost every day, and the price in yuan was about 5 to 10 percent higher than it was on Mt. Gox and Bitstamp (when the exchange rate between the dollar and yuan was taken into account). On Saturday, with everyone still in the office, the price surged again, jumping from 2,100 to 2,500 yuan, or some 20 percent, in the course of a few hours. On the dollar exchanges the price was approaching $400. At the end of a nonstop weekend of work, Bobby sent an e-mail to motivate his staff:

> During the coming days, the market will continue to be super hot, and our workload will be non-stop.
> I urge everyone to stay focused, do our job, and keep up the high quality.
> Once the market cools down, with more normal trading volumes, then we can take a break and evaluate how things go.

Everyone looked for reasons that could explain the continuing rise but, as is often the case in speculative markets, the upward moves seemed to be less dependent on outside events than they were on previous upward moves in the market. Bobby had guessed so many months before that the Chinese would want to bet on something that seemed to have momentum, and Bitcoin's ascent was proving him right.

In the midst of this, Bobby and his cofounders decided to do their part to increase the excitement by making a public announcement about the $5 million investment they had secured back in September and had kept quiet until now. Over the weekend of November 16 and 17, Bobby worked with his investor and a few news sites to prepare an announcement for Monday morning. When the story hit, the already rising price began to move that much faster, rising 15 percent in the course of a few hours, to a price that was already more than twice what it had been at the beginning of the month. But this was to be only the start of a very long day.

●

CHAPTER 26

November 18, 2013

Several hours after Bobby Lee announced the $5 million investment in Shanghai, Patrick Murck, the general counsel of the Bitcoin Foundation, woke up in a hotel room in Washington, DC, and checked on Bitcoin's rising price. After putting on his plain black suit and carefully attaching an American flag pin to his lapel, he left his room, carrying the testimony that he had been writing for the last few weeks and that he was about to present to the United States Senate.

Since appearing at the private meeting with lawmakers back in August, Patrick had spent much of his time helping a staffer for Senator Tom Carper of Delaware, who wanted to hold a hearing on Bitcoin in the Homeland Security and Governmental Affairs Committee. A young aide, John Collins, had gotten excited about Bitcoin earlier in the year and had been holding private conversations across Washington about the technology. When Bitcoin took off in the fall, it helped Collins finally make the hearing happen.

Collins and Patrick shared a similar genial sensibility and a dry sense of humor, and they fell into an easy relationship. Patrick made sure Collins had all his questions answered by the most presentable people in the Bitcoin world, including representatives of all the companies that had won funding from venture capitalists earlier in the year. For the hearing, Patrick's goal was to present the most mainstream image of Bitcoin possible. He volunteered to testify himself, alongside a few other relative newcomers to the Bitcoin world who Patrick knew would say the sorts of things that would make lawmakers happy.

The night before the hearing, Patrick had trouble sleeping and kept rising to make tweaks to his prepared remarks. Patrick also worried about the first part of the hearing, which was a panel of government officials whom he had not been able to prep. Over the summer, Patrick had spoken with all the agencies represented on the panel, but he didn't know if the lower-ranking officials had conveyed his message to their bosses at the Department of Justice and the Secret Service.

When Patrick got to the hearing room and took his seat in the audience for the panel of government officials, he was exhausted and jittery. There were, though, already good headlines trickling out. In response to a questionnaire from Senator Carper's committee, the chairman of the Federal Reserve, Ben Bernanke, had written down his take on Bitcoin and was surprisingly positive, praising its "long-term promise, particularly if the innovations promote a faster, more secure and more efficient payment system."

First up to testify was the head of Financial Crimes Enforcement Network, or FinCen, Jennifer Shasky Calvery, who had helped set up the August meeting. Patrick had developed a good relationship with Shasky Calvery, but she was even more positive than Patrick expected, using his frequent line that cash dollars were actually the most commonly used currency for drug deals and money laundering. The head of the Department of Justice's criminal division went next and emphasized that Bitcoin was not as hard to track as many people seemed to believe and had many legitimate uses. Finally, the head of criminal investigations at the Secret Service said that his agency was not overly worried about its ability to deal with crimes involving virtual currencies.

In response to questions from Senator Carper, the panelists pointed to all the activity in China and noted that if the United States came down too hard on Bitcoin, or pushed it out of the country, the innovation would be likely to move overseas to places like China where it would be harder to control. By the time the first panel was over, the *Washington Post* already had a headline that read "THIS SENATE HEARING IS A BITCOIN LOVEFEST."

When Patrick and the other Bitcoiners got their chance to testify, Patrick was still nervous enough that he forgot to turn on his microphone. But he had a simple message for himself that he repeated over and over: "I've already won, just don't fuck it up. Just read the script."

He didn't fuck it up, and neither did the men sitting next to him. The hearing was streamed live over the Internet and Bitcoiners watching it around the world responded by buying coins and then more coins, pushing the price up as the hearing went on. When Senator Carper brought the gavel down, the price on Mt. Gox stood above $700, $150 higher than where it had been that morning.

Patrick wanted to crawl into bed but first he had to make it through a series of press interviews, including one with a Chinese journalist from CCTV.

THE NEXT MORNING, Bobby Lee arose in Shanghai to discover that BTC China customers had responded with more vigor than even the customers trading dollars on Mt. Gox and Bitstamp, sending the price above 7,000 yuan. In other words, since the previous morning, the Bitcoin price in yuan had gone up more than it had in the first five years of the virtual currency's existence.

Bobby raced to his office, where there was already a journalist from the Xinhua News Agency waiting for an interview. Everyone wanted to know what Bitcoin was and how long this surge could continue.

After the interview, Bobby grabbed Ling Kang, a slight man who had become Bobby's all-around fix-it guy since he came on two months earlier, handling all relations with the government thanks to his incredible connections, or *guanxi* as the Chinese put it. Once they were in the glass-walled conference room behind Bobby's desk, they gave each other dazed looks. They both agreed that the speculative frenzy, which had once been exciting, was now a potential problem. Unlike officials in the United States, Chinese officials had given no encouraging signs about Bitcoin. Also, compared with those in the United States, officials in China tended to act much more swiftly and decisively when they didn't like something. Bobby and his deputy couldn't help recalling how the speculation in Q coins had been shut down. Communist officials now had no shortage of indications that Bitcoin was the new Q coin. A story the previous week in Xinhua had said that even "Chinese mothers" were plowing their money into the virtual currency.

They began talking about what they might do to rein in the excess, including reintroducing trading fees so that buying and selling coins would no longer be free. But other Chinese exchanges had also removed trading commissions and were nipping at BTC China's heels. If Bobby imposed fees, customers would simply flee to the other exchanges. What's more, Bobby and Ling didn't want to give any sign of panicking.

Before they could make any moves, more encouraging news came out of Washington—the last thing Bobby needed. A day after the hearing chaired by Senator Carper, the Senate Banking

Committee had its own hearing on virtual currencies, which covered much of the same territory and drew much less attention. At the end, though, Senator Chuck Schumer, a member of the banking committee, entered the hearing room. This was the man who, back in 2011, had called for a crackdown on Silk Road and implied that Bitcoins were a part of the problem. Now, he wanted to let it be known that he had been misunderstood.

"I do not want to shut down or stamp out Bitcoin," Schumer said. "The potential for a new payment platform and the rise of alternative currencies could have profound and exciting implications for the way we conduct financial transactions."

THE UNMISTAKABLE IRONY of these wild days was that a technology that had been designed, in no small part, to circumvent government power was now becoming largely driven by and dependent on the attitudes of government officials.

This was no accident. Patrick Murck and the new Silicon Valley advocates for Bitcoin had been arguing for months that the technology was not, as Satoshi Nakamoto had initially intended, a network that allowed participants to make anonymous transactions outside the reach of the government. At the Senate hearings, the Bitcoin panelists all emphasized that the virtual currency was actually a terrible way to break the law. With the full record of transactions on the blockchain, the Bitcoin advocates said, it was often possible to identify the people involved in transactions, or at least more possible than it was with transactions involving cash.

But the advocates for the original vision of Bitcoin were not folding their tents and going away. Not long after Ross's arrest, Silk Road 2.0 showed up on the dark web, offering the same services in essentially the same format that Ross had used. The arrests of moderators and administrators from Silk Road 1.0 kept coming, but this wasn't serving as a deterrent. Beyond merely resurrecting the old Silk Road, some developers began trying to devise a truly decentralized online market, which would not have to rely on the sort of centralized escrow service that Ross Ulbricht and his staff had provided and that had ultimately proved to be the site's worst weakness.

Meanwhile, on the Bitcoin forums and Reddit the libertarians and anarchists were more passionate than ever in their defense of the original spirit of Bitcoin and in their criticism of the accommodationists at the Bitcoin Foundation and elsewhere.

Roger had evolved into the spiritual leader of this wing of the Bitcoin community. He had been one of the only people who had chosen not to respond to the inquiries from the Senate committee. In early December Roger used some of his Bitcoin holdings, which had gone up in value thousands of times, to make a $1 million donation to the Foundation for Economic Education, an organization fighting for limited government and free market policies. Roger had also continued to be outspoken in his advocacy of a Bitcoin network that didn't require users to hand over lots of personal information. At Blockchain.info, he supported the development of Shared Coin, a service that mixed up coins from different transactions so that it was impossible to tell which ones came from which addresses. Roger spent most of November in England with the founder of Blockchain.info and his newly hired CEO, looking at ways to expand the company. The number of Blockchain.info wallets had grown to almost 700,000 from 350,000 just a few months earlier. When Roger needed a break from the work, he would visit the local jujitsu dojo with his custom-made kit, or uniform, featuring a big gold Bitcoin emblem on the back.

There were several other programmers and entrepreneurs pushing in a similar direction. Tinkering with the Bitcoin protocol, programmers had created whole new cryptocurrencies, like Anoncoin and Darkcoin, which were explicitly designed to preserve the anonymity of their users. Within Bitcoin, the most ambitious projects aimed to build services that allowed for the exchange of dollars and euros for Bitcoins without going through a central service like Coinbase or Bitstamp. Everyone now saw that any company that handled traditional currencies would inevitably be subject to traditional regulations.

Events in the broader world validated many of the fears that had originally driven the Cypherpunks and Satoshi to imagine a revolutionary new currency. Government documents leaked by Edward Snowden showed, over the course of 2013, that the National Security Agency had indeed been secretly monitoring the electronic communications of a wide swath of American citizens. But the relatively apathetic public response to the tales of NSA surveillance suggested that most Americans didn't actually care much if the government was collecting information about them. What did it matter to the ordinary citizen if he or she wasn't doing anything wrong?

Within the growing Bitcoin community, there was a similar sense that most users weren't all that concerned about the total privacy of their transactions. Perhaps more important, with the price of Bitcoin now hovering near $1,000, there was a growing swell of voices talking about the virtues of Bitcoin that had nothing to do with whether a government could or could not track users.

On December 1 the first-ever research on Bitcoin from a Wall Street firm was released; this report called it a "potentially game-changing disruption" to the payments industry. Gil Luria, a research analyst at the trading firm Wedbush, wrote about the technology with the kind of excitement normally found at Bitcoin meetups.

"We see the intrinsic value of Bitcoin as the conduit in a new global crowd-funded open-source payment network," Luria wrote.

By Luria's analysis, Bitcoin had tapped only 1 percent of its potential market and the price of each coin could easily go up to ten or even a hundred times its current level, to some $100,000 a coin.

The same points got more attention when they were made four days later in a research report from Bank of America Merrill Lynch, the first of the major banks to chime in. Bank of America's chief foreign exchange strategist, David Woo, expressed more notes of skepticism than Luria, pointing to the dangers of Bitcoin's volatility and association with the underworld. But Woo's fourteen-page report noted that in addition to the possibility of a new payment network, Bitcoin could "emerge as a serious competitor" to money-transfer businesses like Western Union.

Woo's price forecast for Bitcoin was not as optimistic as Luria's, but he argued that the services Bitcoin offered could be worth, in total, as much as $15 billion, or $1,300 per coin.

The notion that Bitcoin could provide a new payment network was not terribly new. This is what Charlie Shrem had been talking about back in 2012, and BitPay was already using the network to charge lower transaction fees than the credit card networks. But the idea took on a different weight when it came from employees at banks that had the potential to adopt and popularize the technology.

The clearest indication of how quickly this was moving came not from the public research reports, but instead from an e-mail that Pete Briger, the chairman of Fortress Investment Group, got from a top executive at Wells Fargo, the nation's largest bank by certain measures.

Briger had, in the summer, floated the idea of Fortress partnering with Wells Fargo on a mainstream Bitcoin exchange. Then, the bank had declined to pursue the opportunity and Briger had pulled back on his big ambition to get Fortress into the virtual currency space. Now, though, Wells Fargo was back and wanted to reopen the conversation. The men began planning a meeting at Fortress's New York headquarters. Wells Fargo would never do anything that conflicted with its government regulators, but it now seemed possible to do Bitcoin work with the blessing of those regulators.

WHILE BITCOIN WAS winning mainstream approval in the United States, it was moving in the opposite direction in China. On December 5, just after Bobby Lee had boarded a plane in Shanghai for his first business trip to the United States since Bitcoin had exploded in China, he got a call from a reporter at Bloomberg News, who explained that sources were telling him that China's central bank, the People's Bank of China, was about to release virtual-currency regulations.

This was news to Bobby. The deputy governor of the People's Bank had said back in November, in unscripted comments, that Bitcoin was unlikely to get legitimacy, but that people were nonetheless free to participate in the market. That had led many people to assume that the central bank would take a hands-off approach. This had helped the frantic speculation on Bitcoin to continue, with the price above 7,000 yuan on the day Bobby was flying to San Francisco.

But as a longtime observer of markets, Bobby knew this frenzy was unlikely to end with anything other than a dramatic crash and, when it did crash, it was not going to help Bitcoin's long-term popularity or status with the Chinese government. Bobby had been warning people that the price was unlikely to keep rising, but he wasn't averse to some help from the central bank.

"We're happy to see the government start regulating the Bitcoin exchanges," Bobby told the reporter before quickly signing off.

Bobby spent the flight in an optimistic mood, imagining that the uncertain state in which he'd been operating would soon be cleared up. But when the plane landed and he turned on his phone, he had over a dozen messages waiting for him. In one of them, his head of government relations, Ling Kang, said, "Whatever you do, call me first."

On the long walk to customs, Bobby got Ling on the phone and told him he had heard about the regulations before taking off.

"No, no," Ling said in the Mandarin they used in conversation, with an audible note of fear in his voice. "Bobby, this is the real deal."

The document that had been released while Bobby was in the air was indeed from the People's Bank, but it was also signed by four other major ministries, and it created deep uncertainty for the future of Bitcoin in China, Ling said.

The good news was that the agencies had declared that Bitcoin was not in itself illegal and could be considered a kind of digital asset that people should be allowed to buy and sell. The document also said that virtual-currency exchanges needed to register with the Ministry of Information; this suggested that the exchanges weren't going to be shut down.

The bad news, Ling explained, was that the government had ruled that Bitcoin was not a currency, but was, instead, a digital commodity.

The Chinese government had stepped right into the middle of the ongoing debate about how to define Bitcoin and had actually found itself in agreement with Wences Casares and many other advocates for Bitcoin, who believed that in 2013 the files on the blockchain were more similar to commodities, like gold, than to currencies, like dollars and euros, because Bitcoins were not yet widely or easily used as a medium of exchange or as units for accounting. Beyond those qualities, the Chinese government had also said that Bitcoin lacked the most important characteristic of a currency: government backing.

The Chinese government's categorization of Bitcoin as a digital commodity didn't, on its face, seem terrible to Bobby. Within China, almost no one was using Bitcoin to buy and sell things—it was still just a speculative investment. The problem, though, was that because it was not considered money, the government had declared that banks and payment processors could not deal with Bitcoin, either directly or indirectly.

Bobby grilled Ling on what this meant. Would Tencent, the payment processor, have to stop transferring yuan to BTC China for customers if Tencent itself wasn't touching Bitcoins? If so, that could be deadly.

As was often the case with Chinese government statements, the specifics were left unclear, giving party officials flexibility to deal with the situation as it progressed. Ling wasn't hopeful about where this would lead. The statement made it clear that government officials were not happy with the degree of speculation they had seen.

But Bobby was an American-educated optimist and Tencent hadn't shut BTC China down yet. What's more, there was obvious room in the statement for them to continue doing business.

The market seemed to agree with Bobby. In the hour immediately after the Chinese government statement had come out, the price of Bitcoin had entered a free fall, dropping 25 percent to 5,200 yuan. But soon thereafter the price began recovering, and it was already back to around 6,400 by the time Bobby was through customs.

That afternoon Bobby gave a talk at his alma mater, Stanford, and explained that he was "cautiously optimistic" about the new rules. But that day's statement was not the final word from the government.

CHAPTER 27

●

December 7, 2013

The extent to which Bitcoin could survive and grow without government approval was on display in Buenos Aires, at the first conference hosted by Bitcoin Argentina. The group had been founded by Wences Casares's old friend Diego along with a partner he had met at a Bitcoin Meetup earlier in the year. For the conference the men had booked a big hotel in downtown Buenos Aires and managed to sell four hundred tickets, with about 40 percent going to foreigners like Roger Ver, Erik Voorhees, and Charlie Shrem.

The ticket-buying process itself had put a spotlight on one of the most promising Bitcoin startups to emerge from Argentina and one of the first companies anywhere using the network to legally provide a service that wasn't possible with the traditional financial system.

In Argentina, credit card transactions with foreigners, like the sale of conference tickets to Americans, normally took a long and expensive route before paying out in Argentina. The American customer's credit card company would deduct around $2.50 from the $100 ticket price to send the money to Diego's Argentinian bank. From there, the Argentinian bank would generally charge another 3 percent for the foreign exchange, leaving $94.50. The big hit, though, happened when the Argentinian bank turned the dollars into pesos. If Diego converted the $94.50 with a money changer on the street he could have gotten the unofficial rate of around 9.7 pesos for each dollar, leaving him with 915 pesos. But the bank exchanged the money at the official exchange rate set by the government—6.3 pesos at the time of the conference—giving him, instead, 595 pesos. On top of that, Diego's bank wouldn't give him those pesos until twenty days after the customer purchased the ticket.

The Argentinian Bitcoin startup, BitPagos, provided a clever way around this expensive morass. BitPagos took the $100 credit card payment in the United States and charged a 5 percent fee. But instead of transferring the remaining $95 to an Argentinian bank, BitPagos used the dollars to buy Bitcoins in the United States. Bit-Pagos then transferred the Bitcoins directly to Diego. He could either keep the Bitcoins or exchange them for pesos at the unofficial exchange rate, thus ending up with around 920 pesos, instead of 595. And rather than taking twenty days, BitPagos gave him his Bitcoins in two days.

BitPagos had been started earlier in the year by two young Argentinians, a man and a woman, who had been running a consulting company and struggling to take payments from foreign customers. In addition to collecting ticket payments for the foundation, the new company was getting traction with hotels that took money from foreign tourists and didn't want to pay the cost of getting those payments into pesos. By the time of the conference, BitPagos had already signed up around thirty hotels. Most of these hoteliers didn't care about the ideas behind a decentralized currency; they were just happy to find a way around the expensive tollbooths that littered the Argentinian financial system. As an added bonus, they could end up with money in Bitcoins rather than the rapidly depreciating peso.

This was an eminently practical use of Bitcoin to deal with the inflationary mess in Argentina, but it was so practical that it actually swung around into the domain of the ideological ambitions that Satoshi Nakamoto and the Cypherpunks had imagined. The Argentinian hoteliers might not have been libertarians, but they would have easily understood Satoshi's early writing about Bitcoin, which explained that "the root problem with conventional currency is all the trust that's required to make it work. The central bank must be trusted not to debase the currency, but the history of fiat currencies is full of breaches of that trust." Mismanagement of currencies was a part of daily life in Argentina.

The conference in Argentina attracted many of the more ideologically minded Bitcoin followers from around the world. The old team from BitInstant gathered for a reunion of sorts and the team members were all given prominent speaking spots. They lived it up in Buenos Aires, eating steak, drinking Argentinian wine, and going to a tango performance with the other presenters at the conference. But for them and most of the foreigners at the conference, the most memorable thing about the event was not a part of the official proceedings. Everyone coming into the Hotel Melia, where the conference was held, passed two teenagers, a boy and a girl, whose wispy, almost ethereal features gave them away as twins. Both of them were wearing the same white T-shirt with the word *Digicoins* on the front, and both asked people entering the conference, in a gentle voice, if they wanted to buy or sell Bitcoins. Those who took them up on the offer were guided to a Subway sandwich shop across the street. There at a table sat a man with wavy silver hair, dark eyes, a computer, a white shirt unbuttoned enough to reveal his chest hair, and a backpack full of cash.

The man, the father of the twins, had his Bitcoin wallet up on the laptop and he could change money in either direction, in much the same unofficial way as all the other black-market money changers on Buenos Aires street corners. Dante Castiglione, the owner of Digicoins, had not created Digicoins just for this conference. He had, by this time, been serving as one of Argentina's most successful virtual-currency exchangers for a few months. His twins were his runners, going out into the city each day to visit the customers in need of pesos or virtual currency. When people asked about his business, he was stingy with details and gave a wry smile, as if to ask, "Why do you think I'm doing this?" But he was willing to say that this was only the latest stop in an itinerant career built by finding opportunities in Argentina's broken financial system.

"I am a working man," he would say when pushed. "We are trying to give our service. We are earning our food and our rent."

Bitcoin's evolution in the United States and China was showing how the technology could become dependent on the official financial system and government approval. Argentina, on the other hand, was showing how it could develop without any of that. It certainly moved more slowly, but there was something more tangible and grounded about what was being created.

THE MAN WHO had gotten this ball rolling in Argentina, Wences, couldn't make it to the conference in Buenos Aires. At the time, he was finalizing the sale of his most recent startup, Lemon, for $42.6 million. When he wasn't winding down his work with Lemon, he was working on the new Bitcoin company he was creating with Fede Murrone, his longtime collaborator in Argentina.

The core of the new business was the system that Wences and Fede had begun developing early in the year to store their own, significant holdings of Bitcoin, having come to distrust Mt. Gox and the other available services. Their main goal had been to get the private keys for all their addresses off any computer hooked up to the Internet. Wences and Fede had begun by putting their private keys on an offline laptop and storing that laptop in a safe-deposit box at a bank in California; this allowed them to delete all the private keys from their online computers.

Over the course of 2013, the value of their Bitcoins had grown, as had the number of people who heard about their system and asked to store Bitcoins on the laptop. This had provoked Wences and Fede to take more and more strenuous measures to secure the private keys. First, they encrypted all the information on the laptop so that if someone got hold of the laptop that person still wouldn't be able to get the secret keys. They put the keys for decrypting the laptop in a bank near Fede in Buenos Aires. Then they moved the laptop from a safe-deposit box to a secure data center in Kansas City. By this time, the laptop was holding the coins of Wences, Fede, David Marcus, Pete Briger, and several other friends. The private keys on the laptop were worth tens of millions of dollars.

The interest shown by friends suggested to Wences that there was a broader need for a more reliable way to store Bitcoins. People didn't want to hold the private keys on their home computers, but they also didn't trust Mt. Gox and Coinbase to keep digital files worth millions. The vault, as Wences and Fede called it, was just a starting point. Wences imagined that this would be the first offering in what would become a full-service Bitcoin company that could provide a place for people everywhere to store and spend their coins. Unlike the previous startups that Wences had started and sold, this one was intended to be his lifework—the last company he would ever found. He called it Xapo, a name that he and Fede settled on after looking for a simple, distinctive word for which the dot-com domain name was available.

Wences initially had little interest in taking money from investors for this company. He didn't want to give control to anyone else and he had enough money to pay for it all himself. But over the fall of 2013, his friends convinced him that starting a company without investors would deprive him of all the connections and marketing possibilities that funders bring.

The value of having investors became very clear to Wences the same day that he completed the sale of Lemon, when Coinbase announced that it had raised $25 million from Andreessen Horowitz to grow the company. It was the biggest public investment in a Bitcoin company, by a good margin, and Coinbase reaped the reward in new customers and attention.

A few days after this, Wences journeyed to San Francisco to meet with Benchmark, a venture-capital firm that had been vying with Andreessen Horowitz to invest in Coinbase. Wences had been friendly with the Benchmark partners for some time, and he had hoped he might find an opportunity to work with them. One of them was the brother-in-law of Fortress's Pete Briger.

The meeting at Benchmark's offices was unlike Wences's earlier fund-raising efforts. This time, he laid out what he needed from Benchmark to make it worth his while. After Wences's presentation, the Benchmark team huddled briefly and then offered to put $10 million into Wences's company, valuing it at $50 million. As in all of Wences's past startups, there was no term sheet, just a handshake.

When Wences walked out, he immediately called his old friend Micky Malka to tell him the exciting news. Micky responded not with excitement, but instead with pique, because Wences hadn't offered Micky and his firm, Ribbit, a place in the deal. After demanding an opportunity to put $10 million into Wences's company, Micky finally settled for $5 million. A short while after that,

Pete Briger called to demand a place in the round too, and Wences agreed to let him put in $5 million. This left Wences with $20 million before he even had a functioning business.

DURING HIS TWO-WEEK stay in the United States, Bobby Lee visited his brother Charlie, who had quit his job at Google over the summer and joined Coinbase to work on Bitcoin full-time. Bobby showed up at the company's makeshift offices in a converted three-bedroom apartment a day after the company announced the $25 million investment from Andreessen Horowitz.

Charlie Lee didn't need to work another day of his life. Litecoin, his alternative cryptocurrency, which was a slightly faster, lightweight version of Bitcoin, had now become the second-most-popular cryptocurrency in what was becoming an increasingly crowded field of Bitcoin knockoffs. In part because of Charlie's transparency in launching Litecoin, people trusted it and were betting that it would be, as Charlie had intended, the silver to Bitcoin's gold.

In November the value of all the outstanding Litecoins had briefly surpassed $1 billion. The particular computer chips that were good for mining Litecoins were sold out at nearly every online electronics retailer. Charlie had been mining Litecoins since the beginning, so he owned a sizable number of the coins, along with his significant Bitcoin holdings. His work at Coinbase was primarily due to his desire to help bring virtual currencies into the mainstream.

Charlie saw that Bitcoin had done similarly good things for Bobby. Despite all the long hours and uncertainty Bobby had endured over the last few months, his position as a CEO, after years in middle management, had given him a confidence and self-assurance that seemed to outweigh the stresses of the job.

Bobby had spent much of his time in the United States looking for new investors and partners for BTC China. But he was still trying to figure out what the People's Bank of China statement on December 5 would mean for his company moving forward. On Bobby's exchange, the price of Bitcoin had fallen from the all-time highs, but it stabilized at around 5,500 yuan, or $875, on Western exchanges. Bobby learned from his staff that the December 5 statement had come about after the enormous price spike in November. Several reports had gone up to the State Council, the highest administrative authority in China, and one of the four vice premiers of the council had ordered the People's Bank to do something about the situation. As is generally the case in China, the whole process was enshrouded in secrecy and seemingly driven by officials trying to protect their backs.

On Bobby's last night in the United States, his government-relations guru, Ling Kang, called again. The payment processor Tencent had just called BTC China to explain that Tencent was going to stop doing business with Bobby's exchange in the next few days. Bobby was furious. Tencent had previously agreed to provide at least a ten-day notice of any changes. That night, he called everyone he could think of to argue his case. But he and Ling heard back that Tencent had gotten orders directly from the local branch of the People's Bank and there was no fighting it.

When Bobby flew back to China the next day, everyone at his company was scrambling to get a new payment processor set up before Tencent shut the company off on Sunday at noon. But it now appeared that the problem wouldn't end with Tencent. Bobby learned that all the payment processors had been called to the People's Bank on Monday to discuss the issue.

The Monday meeting did not generate any official change in policy or new documents. But the real-time reports from the meeting that Bobby's team was receiving revealed that the payment processors were all being encouraged to reconsider any business with Bitcoin companies. As the rumors began to leak, the price dropped, falling to around $600 on Western exchanges. Two days later, when Bobby officially confirmed that his company would stop taking new deposits, a new sell-off began, taking the price down to $430 on Bitstamp and 2,100 yuan on BTC China, or less than a third of what it had been at the high just two weeks earlier. Whereas 100,000 Bitcoins had been trading hands daily on BTC China a few weeks earlier, now the trading volumes were less than a tenth of that.

Bobby was in back-to-back meetings with his staff contemplating ways to stay alive without the payment processors. One of the other Chinese exchanges, Huobi, began taking in customers' money through the personal bank account of the company's CEO. The December guidance from the Chinese central bank seemed to bar banks from working with Bitcoin, but Bobby was surprised to see that the banks eagerly took the business from his competitors. Bobby's Chinese deputies explained that the banks were doing this because, unlike the payment processors, they had not been called into a meeting and warned not to work with Bitcoin. Whereas in the United States, banks were unwilling to do work unless they were explicitly given a green light by regulators—and sometimes not even then—in the Wild West of China, the banks would try just about anything until they were explicitly told it was not allowed.

Bobby, though, had worked most of his adult life for American companies and he was uncomfortable skirting the rules. The best alternative seemed to be some sort of voucher system, in which third-party vendors would sell credit for BTC China, similar to the way vendors sell cards with cell phone minutes. But as his staff rushed to get this set up, Bobby watched customers flock to the competitors who had set up bank accounts. In China, scrupulously following the rules seemed to be a recipe for losing business.

•

EACH NEW RUN-UP in the price had drawn new and more sophisticated scrutiny of the principles underlying Bitcoin, and the
December rise and fall were no different. This time, the people
training their sights on Bitcoin were some of the highest-profile
economists in the United States—including Paul Krugman, the
progressive Nobel Prize winner; and Tyler Cowen, the prolific
libertarian-leaning blogger. Few of them had much good to say.

Krugman focused largely on Bitcoin's claim to be a currency,
given the difficulty it seemed to have fulfilling one of the basic
roles of money: serving as a reliable store of value. Why would
people store their wealth in Bitcoin if they knew the value was
going to fluctuate so violently? Krugman asked.

Cowen, meanwhile, argued that Bitcoin was going to have
difficulty sustaining its value as new and better-designed cryptocurrencies came along and drew users away from it. Some people
were, indeed, already choosing to hold Litecoin, Charlie Lee's creation, and a hip, younger cryptocurrency, Dogecoin.

But a deeper strain lurking beneath these critiques was an
awareness that one of the fundamental premises that had driven
Bitcoin's popularity seemed, increasingly, to have been disproved.
Many early Bitcoiners, particularly in the libertarian camp, had
believed that the Federal Reserve's efforts to stimulate the economy in the wake of the financial crisis, by pumping lots of new
money into banks, would devalue the dollar and lead to high inflation, similar to what had happened in Argentina.

This idea made a scarce asset like Bitcoin or gold look like a
safer bet than holding dollars. But in late 2013 none of the fears
about inflation had been borne out. In fact, the problem facing the
American economy was not inflation, but deflation, because banks
were holding much of the new money, rather than putting it out

into the economy. The Fed's stimulus program had been successful enough that the European and Japanese central banks were now copying it. This was a living economics experiment and it didn't seem to be going the way libertarians expected. At the same time, the scarcity of Bitcoins still had the effect that early critics had warned about: it was encouraging people to hoard Bitcoins rather than actually use them.

Perhaps the most stinging criticism came from a well-known British science fiction writer, Charlie Stross, who wrote out a long list of Bitcoin's potentially damaging effects, of which some were intended by the Cypherpunks (for example, tax evasion and weakening government social-welfare programs) and some were not. Stross noted that in the latter category, the hoarding encouraged by Bitcoins' scarcity was leading to a vast inequality in the holdings of Bitcoins, "to an extent that makes a sub-Saharan African kleptocracy look like a socialist utopia." Indeed, a few Bitcoin holders, like Roger Ver and Wences Casares, owned a material proportion of all the outstanding coins. This was unlikely to sit well with the Occupy Wall Street crowd, who objected to the undue power of the wealthiest 1 percent of the population.

The Bitcoiners had their ready responses to all these critiques and voiced them loudly. Bitcoin's volatility would go away as it matured, the believers said, and Bitcoin had a first mover advantage against other cryptocurrencies that was showing no signs of abating. Meanwhile, inflation might not be a problem in the United States yet, but it was a problem in other countries.

Whatever the merits of the criticisms, they did not seem to be dulling the growing curiosity about Bitcoin within major financial institutions. The most notable name to show signs of interest was Wells Fargo, perhaps the nation's most successful and most respected bank in the wake of the financial crisis. After the Senate hearings in November, Wells Fargo executives had reached out to

Pete Briger to reopen the conversation about working together on a Bitcoin exchange. One sign of Wells Fargo's openness was that executives of the bank agreed to travel to Fortress's headquarters in New York for the meeting. Briger rounded up a team of people to make the case for Fortress, one that included Wences and others who flew in from California.

Fortress put aside a grand conference room on the forty-seventh floor of its Manhattan headquarters, and executives from several divisions of Wells Fargo showed up. Once the dozen or so people were gathered around the conference room, Pete stood up and made his basic pitch to the Wells Fargo team. He explained why the Fortress team was so intrigued by the technology and pointed at the smart people around the table, such as Wences, who had thrown themselves into it. He hinted that Wells Fargo should be keeping up with Bitcoin, given the potential for the new network to challenge some of the basic services, like payment networks, that the bank was providing. Pete closed by talking about the lack of an American-based regulated exchange for Bitcoin—something that Fortress and Wells Fargo could provide together.

The questions from the Wells Fargo executives did not reveal much about how serious the bank was about the project, but they had clearly done their homework and came with detailed questions about what exactly an exchange would look like and how it might satisfy regulators. The meeting concluded with an understanding that the bank would take it all under consideration.

The potential advantages of Bitcoin over the existing system were underscored in late December, when it was revealed that hackers had breached the payment systems of the retail giant Target and made off with the credit card information of some 70 million Americans, from every bank and credit card issuer in the country. This brought attention to an issue that Bitcoiners had long been talking about: the relative lack of privacy afforded by traditional payment systems. When Target customers swiped their credit cards at a register, they handed over their account number and expiration date. For online purchases Target also had to gather the addresses and ZIP codes of customers, to verify transactions. If the customers had been using Bitcoin, they could have sent along their payments without giving Target any personal information at all.

During this period, it was notable that some of the most encouragingly positive statements about virtual currencies came out of branches of the Federal Reserve, the archetype of the central bank that Bitcoin had set out to supplant. Fed officials didn't love the idea of a currency outside the control of governments, but they were very eager to see methods of moving money that cut out middlemen, who introduced risk into each transaction and into the financial system. The Fed had, in fact, been making increasingly vocal calls for technology that would allow more direct methods of moving money. During late 2013 and early 2014, a number of branches of the Federal Reserve put out papers discussing the potential for the blockchain technology to eliminate risk in the financial system, if this technology could be harnessed properly.

"It represents a remarkable conceptual and technical achievement, which may well be used by existing financial institutions (which could issue their own Bitcoins) or even by governments themselves," a Bitcoin primer released in late 2013 by the Federal Reserve Bank of Chicago said.

Bitcoin's use as a new, more secure, and more private way to make payments online was given a big boost in early January 2014 when the online retailer Overstock announced that it would begin accepting Bitcoin for all purchases. The eccentric chief executive of Overstock, Patrick Byrne, had a PhD in philosophy from Stanford and was an outspoken libertarian. He clearly had political motivations for taking Bitcoin, hoping to get the country out from "under the thumb of Wall Street oligarchs," as he put it. He also pointed to all the eager Bitcoiners looking to spend their money with anyone who would take the currency. But in interviews he emphasized the more practical reasons for any company to make the move: no more paying the credit card companies 2.5 percent of each transaction (the company helping Overstock take Bitcoin, Coinbase, charged Overstock 1 percent); no more dealing with chargebacks from customers who received shipments and then disputed the charges; and no more worrying about holding lots of sensitive financial information for customers. On the first day, Overstock processed more than $100,000 in orders paid for with Bitcoins.

CHAPTER 28

●

January 20, 2014

Wences Casares pulled his white Subaru Outback into an elegant, understated strip mall off Woodside Avenue, one of the main roads winding down out of the hills above Palo Alto. It was 7:30 a.m. and Wences was looking forward to his breakfast at Woodside Bakery and Café, a favorite spot for Silicon Valley deal making that provided a bit more seclusion than the restaurants down in Palo Alto.

The man waiting for him inside was often referred to as the best-connected person in the Valley, and not just because he had cofounded LinkedIn, the business-networking site. Reid Hoffman's girth and bearing hinted at his larger-than-life character. After studying at Oxford, on an elite Marshall Scholarship, Hoffman had been brought in by Pete Thiel to help build PayPal—Thiel called him the "firefighter-in-chief." Hoffman later introduced Thiel to Mark Zuckerberg, an introduction that led to Thiel's making the first major investment in Facebook. By that point, Hoffman had already begun building LinkedIn with some colleagues from an earlier startup. When Wences first met Hoffman, not long after arriving in Silicon Valley, Hoffman was looking for new investments and serving on the boards of startups. The breakfast at Woodside Café was one of their periodic check-ins. Wences was finalizing the investments in his new company, Xapo, and was eager to tell Hoffman about his plans.

Wences knew that Hoffman had first gotten hooked on Bitcoin by Charlie Songhurst, who had, in turn, gotten hooked on Bitcoin by Wences at the Allen & Co. event in 2013. Hoffman, an expert on social networks, had been captivated by Songhurst's arguments about the power of the incentives built into Bitcoin—primarily through the mining process—that encouraged new users to join the decentralized network while also encouraging powerful miners to do what was best for the system so as not to see their holdings lose value.

"That's actually super important," Hoffman would say later. "That makes it less of a pure technological marvel and more of a potential social movement."

But Hoffman had remained skeptical and was particularly put off by the suggestion that Bitcoin would replace credit cards—the possibility that all the bank research reports were talking about. Credit cards seemed to work pretty well in Hoffman's estimation. Despite the security risks and costs to merchants, he didn't see too many consumers complaining about their credit cards failing them. If that wouldn't get people using this new kind of network, Hoffman wondered, what would?

Hoffman had finally gotten a satisfying answer to this at a dinner with Wences and David Marcus and a few other Valley power players late in 2013. Wences agreed with Hoffman that Bitcoin was unlikely to catch on as a payment method anytime soon. But for now, Wences believed that Bitcoin would first gain popularity as a globally available asset, similar to gold. Like gold, which was also not used in everyday transactions, Bitcoin's value was as a digital asset where people could store wealth.

This was enough to get Hoffman to go home from that dinner and ask his wealth adviser—the Valley's most prominent money manager, Divesh Makan—to buy some Bitcoins for his portfolio. When Wences sat down for breakfast with Hoffman at Woodside Café in January, Wences told him about the progress he was making with Xapo.

"Just to be clear, I'd be super interested in investing," Hoffman told Wences.

Wences paused, a bit chagrined.

"I wish you'd told me that the last time we talked," he said.

"You told me you weren't interested in venture investing," Hoffman shot back.

Wences explained that things had changed since they last talked, and that he had decided to take on investors and had struck a deal with Benchmark Capital.

"I just don't think I can include you in that," Wences said. "It wouldn't be the honorable thing to do."

Hoffman was not so easily deterred. He told Wences he was going home to figure out a way they could make it work. Wences said he would do the same.

Hoffman's newfound enthusiasm was part of a broader passion sweeping Silicon Valley in early 2014. While Wall Street research reports were talking about the possibility of a new payment system, the best minds in the Valley were thinking in much more ambitious terms after looking deeply at the code underlying Bitcoin. These views were crystallized, and projected to a much broader audience, the day after Wences's breakfast with Hoffman, when Marc Andreessen, cofounder of the investment firm that had put $25 million into Coinbase, published a lengthy cri de coeur on the *New York Times* website, explaining what had the Valley so worked up.

"The gulf between what the press and many regular people believe Bitcoin is, and what a growing critical mass of technologists believe Bitcoin is, remains enormous," Andreessen explained.

Andreessen's list of the potential uses for the technology was lengthy. It was an improvement on existing payment networks, owing to its security and low fees, but it was also a new way for migrants to move money internationally, as well as a way to provide financial services to people whom banks had left behind. Like many Valley firms, Andreessen's was thinking about intelligent robots, and Bitcoin seemed like a perfect medium of exchange for two machines that needed to pay each other for services.

Beyond all that, though, the decentralized ledger underlying Bitcoin was a fundamentally new kind of network—like the Internet—with possibilities that still hadn't been dreamed up, Andreessen said. He went on:

> Far from a mere libertarian fairy tale or a simple Silicon
> Valley exercise in hype, Bitcoin offers a sweeping vista of
> opportunity to reimagine how the financial system can and
> should work in the Internet era, and a catalyst to reshape
> that system in ways that are more powerful for individuals
> and businesses alike.

Less than a year earlier, Wences had sat in Arizona with Chris Dixon, a young partner at Andreessen Horowitz who had been trying to get the firm to dive into Bitcoin. Now Andreessen himself was becoming the most outspoken public advocate for the technology, taking on a role that had previously been occupied by people like Roger Ver and Hal Finney.

Andreessen had quietly begun his investing in Bitcoin a year earlier, when he put some of his own money into the Series A fund-raising round of the secretive Bitcoin mining company, 21e6, created by the Stanford wunderkind Balaji Srinivasan. Since then, in addition to the $25 million that Andreessen Horowitz had put into Coinbase, the firm had also made a secret $25 million investment in the confidential Series B round for Balaji's mining company. That Series B also included another $10 million from other Series A investors and $30 million more in venture debt. The best-funded company in the Bitcoin world, with $70 million, was one that only a small elite even knew about. Andreessen liked the investment in part because while he and many others in the Valley believed that venture capital firms should not buy Bitcoins outright, he thought it was kosher to invest in a mining company like 21e6 that paid out its dividends in the virtual currency it mined.

Balaji's mining company had already started rolling out its custom-fabricated mining chips in the fall of 2013 and had quickly come to account for 3 to 4 percent of the hashing power on the entire network. In early 2014 the company was planning to pay the first dividends to investors and was building its own dedicated data center that would hold more than nine thousand machines containing the company's custom chips.

Balaji's promise was so great that in late 2013 Andreessen had invited him to become the ninth partner at Andreessen Horowitz, in no small part to help scout out new investments related to virtual currencies and the blockchain. Balaji was as ambitious and utopian as anyone out there about what Bitcoin could do. He believed that it could help open the door for what would essentially be new breakaway countries, created by people wanting to push technological experimentation to the limits.

For Wences, the more immediate indication of how quickly this was all moving came in an e-mail from Hoffman not long after their breakfast. Hoffman had talked with a friend at the venture capital firm Index Ventures, and together they were prepared to offer Wences another $20 million for Xapo. He could still take the $20 million he already had as a Series A, but this could be a quick follow-on—a Series A1. And while Wences's first investors had valued Xapo at $50 million, Hoffman and his partner were ready to value it at $100 million. In little more than a month, Wences had doubled the value of his company.

STANDING BEHIND THE black bar, Charlie Shrem opened a fridge under the liquor and pulled out two beers, a Blue Moon for himself and an Amstel Light for Nic Cary, the chief executive of Roger Ver's company Blockchain.info, who was in New York on a business trip. The bar, EVR, was closed, but Charlie lived right upstairs and had all-hours access thanks to his investment a year earlier. His girlfriend Courtney, who now lived with him, stopped by to see if Charlie needed anything.

Charlie looked noticeably more weathered than he had the previous summer when he shut down the BitInstant site. He had shaved off his youthful curls and grown a scruffy beard that matched his bushy eyebrows. None of this, though, signaled defeat. Charlie was, in fact, benefiting as much as anyone from the rising interest in Bitcoin. He had taken on a role as an unofficial money changer for some of the big holders of Bitcoin, allowing them to sell large blocks of coins without going on an exchange, where big sales could move the price.

More important, Charlie had managed to connect with a new group of investors who were looking at putting up money so that Charlie could reopen BitInstant. The potential investment was a complicated deal, providing a way to pay off the legal bills from the previous summer while also giving the site a more simple regulatory structure moving forward.

After taking a swig from his beer, Charlie boasted that one of the consultants who had been helping him—one who was a former regulator—had told him: "You and some of your friends have become such super experts in finance, law, and Patriot Act and all these things. There are people who have like thirty graduate degrees who don't know as much as you do."

"And I'm like, 'It's Bitcoin,' " Charlie said with a grin.

David Azar, his old investor, was ready to sign off on the deal to reincarnate BitInstant. The one hitch was the Winklevoss twins. Charlie had offered to give the new investors more than half of his own equity in the company—bringing him from a 27 percent stake down to a 12 percent stake. All the twins and David had to do was give the new investors 2 percent of their 25 percent stake. When the twins shot back a curt e-mail dismissing Charlie's offer, Charlie quickly replied that he would provide all the shares to the new investors so that David and the twins did not have to dilute their stake in the company at all. When Charlie met with Nic, he was still waiting to hear back from the twins.

In the meantime, though, Charlie was not twiddling his thumbs. Earlier the same day, he and his girlfriend Courtney had lunch with a few guys who wanted to sell shares in private jets for Bitcoin.

"It's fucking, excuse my language, it's an amazing idea," Charlie said. A few weeks earlier, he had splurged and sold some of his Bitcoins to pay for a private jet to take him and Courtney to the Bahamas.

He also was still working with the Bitcoin Foundation, preparing for its second annual conference, this one in Amsterdam.

"We're looking for a celebrity speaker," Charlie told Nic. "I want to get like Snoop Dogg to come."

"How about Richard Branson?" Nic asked, referring to the mogul who had recently announced that he would be accepting Bitcoin for tickets on Virgin Galactic, his commercial space company.

"A lot of these guys aren't even out of reach," Charlie said.

A few days after seeing Nic, Charlie and Courtney flew to Amsterdam. They stopped by the convention center where the foundation's conference would be held. But the main purpose of the trip was a technology conference in Utrecht that had paid Charlie $20,000 to speak about Bitcoin. Flying home from the gig, in business class, Charlie couldn't help feeling that, after all his earlier struggles, things were starting to work out again.

After landing in New York, he had just presented his passport to the customs officer when another agent appeared, seemingly out of nowhere, and said, "Mr. Shrem, come with us." When Charlie asked why, the agent said, simply, "We'll explain everything," and led him to a holding room. The agent there handed Charlie a warrant for his arrest and told him he was facing charges of money laundering, unlicensed money transmission, and failure to report suspicious transactions.

When Charlie asked for more information he was told the agents would be happy to tell him more if he'd just answer a few of their questions. He knew better than to talk without a lawyer present and so he was left not knowing what conduct had led to the charges. He was allowed into a larger holding room, where Courtney was waiting, crying hysterically. He calmly told her to call the lawyer who had been working on BitInstant and not to answer any questions the federal agents might ask her. While he was talking to her, he was put in cuffs and led away to a black SUV, which took off in a caravan of police cars and traveled to the Drug Enforcement Administration headquarters in downtown Manhattan. After getting booked, Charlie was taken to the Metropolitan Correctional Center, where he was changed into an orange jumpsuit and locked up in a cell by himself. He had the rest of the night to cry and nervously think through all the things that might have gotten him here and all the ways it might play out.

In the morning, the marshals took him to a holding cell under the federal courthouse, where he met with one of the lawyers he had worked with at BitInstant, whom Courtney had called. He learned, finally, that the charges stemmed from his work in early 2012, selling Bitcoins to BTC King, the money changer who had helped Silk Road customers secure Bitcoins to buy drugs. The prosecutors had e-mails in which Charlie acknowledged knowing what the coins were being used for and doing it anyway without filing any suspicious-activity reports with regulators.

Charlie's lawyer explained the basics. The lawyer had reached Charlie's parents and they were ready to put up their house in Brooklyn as collateral for the $1 million bail. But they had conditions: he had to apologize to them and break up with Courtney. When Charlie resisted the conditions, his lawyer told him that he needed to bite the bullet and do what it took to get out.

Once he was released, with an electronic ankle bracelet on, Charlie found his parents and Courtney in the courthouse hallway. They had never met before and clearly had not been talking. When he asked his parents if Courtney could come home with them, they reiterated that if he wanted to be with Courtney they would rescind the bail and he would go back to jail. He privately told Courtney, who was weeping, that he would try to figure something out and call her later. Outside, he climbed into his parents' black Lexus SUV and headed toward his childhood home.

While Charlie had been sitting in the courthouse, the United States attorney in Manhattan, Preet Bharara, the most powerful prosecutor in the country and the same man who had filed charges against Ross Ulbricht four months earlier, publicly announced that his office had unsealed criminal charges against Charlie and the

Florida man known as BTC King, Robert Faiella. At a press conference, Bharara said: "If you want to develop a virtual currency or a virtual currency exchange business, knock yourself out. But you have to follow the rules. All of them."

Charlie's offense was not of the magnitude that usually caused a federal prosecutor to hold a press conference, but Bharara clearly wanted to make a statement that he was taking a close look at virtual currencies.

THE DAY AFTER Charlie's release, and less than a mile from where he'd been in jail, the Winklevoss twins stepped out of a black car in downtown Manhattan to testify at the latest government hearing about Bitcoin. This one was being held in the somewhat run-down offices of New York State's top financial regulator, Benjamin Lawsky, who had subpoenaed all the major Bitcoin companies and investors back in the summer of 2013. Lawsky had previously worked in Bharara's office. The arrest of Charlie and Bharara's press conference, just a day before Lawsky's hearing, looked to many Bitcoiners like a piece of political theater, designed to give Lawsky an excuse for a more vigorous crackdown on the industry.

The hearing itself couldn't help being colored by Charlie's arrest. In addition to the Winklevoss twins, Barry Silbert, who had wanted to invest in Charlie back in 2012, was there to testify, as was Fred Wilson, the respected venture capitalist who had a number of run-ins with Charlie over the years. The only panelist with no tie to Charlie was Jeremy Liew, the California-based venture capitalist who had put money into Bobby Lee and BTC China.

The people who had been invited to appear on the panel showed that since the Senate hearing three months earlier, the center of influence within the Bitcoin community had shifted toward Silicon

Valley and away from the Bitcoin Foundation that Charlie had helped create.

When Lawsky, in his first round of questions, asked about Charlie's arrest, none of the panelists came to Charlie's defense. The Winklevoss twins had released a statement the previous day suggesting that they had been betrayed by Charlie's behavior. Both Wilson and Liew emphasized that Charlie was part of an early Bitcoin community, in which the seeming anonymity of the technology was the most attractive quality.

"It turns out that the market of radical libertarians is not very big," Liew said in his Australian accent.

The diminishing interest in anonymity and central banks did not mean that the panelists had modest ambitions for Bitcoin. They talked about how this new form of money—and the ledger on which it ran—could allow for new kinds of stock exchanges and other things that hadn't even been thought of yet.

"When you are offering free, radically reduced transactions costs, and when you are offering the ability for programmable money that can put a lot of additional functionality on money, then you are talking about a market size of everybody in the world," Liew said.

All the panelists compared Bitcoin in its current form to the Internet in 1992 or 1993, before the first web browser. Back then, there had been lots of excitement in a small circle of technologists about what the Internet protocol could do, but the programs and infrastructure did not yet exist to make it accessible to ordinary people. It had, at the time, been dominated by fringe communities willing to try out untested technology. In 2014, similarly, the Bitcoin protocol wasn't being used in any particularly compelling way, but that didn't mean it wouldn't be in the future once people discovered customer-friendly ways to harness it.

"We are at the beginning of an exciting time, not just for investors but for all of society," Wilson said.

As the hearing went on, it became increasingly clear that Lawsky and the two deputies who were helping him ask questions were eager to work with, rather than against, their panelists.

"A lot of people initially react to something new like this with immediate skepticism. All of us should resist being overtaken by that urge," Lawsky said. "We want to make sure we don't clip the wings of a fledgling technology before it ever gets off the ground. We want to make certain that New York remains a hub for innovation."

Lawsky was a boyish figure with big, attention-grabbing ambitions. In late 2013 he had announced his plans to create what he called a BitLicense for virtual-currency companies. At the hearing he appeared less the hard-edged interrogator and more the slightly nerdy kid trying to get in with the cool tech kids. If nothing else, it was evident that he thought this was an interesting enough technology that he did not want New York to be left out as it developed.

"We need to think internally about how we can be a more modern digital regulator," he said. "It's not simply what our rules are, it's also who we employ, how quickly we act. There's a lot to do."

While the Bitcoin community seemed to have made significant headway with regulators, it was having less success with the banks, particularly after Charlie's arrest.

"Not good" was the simple message that Patrick Murck got, in an e-mail, on the day that Charlie's arrest was announced, from a contact at Wells Fargo who had been eager for the bank to work with virtual-currency companies.

Charlie resigned from his position as vice chairman of the Bitcoin Foundation on the same day as the hearing in New York, but that didn't help. Another executive at Wells Fargo let Pete Briger know that the bank would not be able to move forward with the joint project with Fortress.

Even before Charlie's arrest, there had been indications that the openness that the banks had exhibited toward Bitcoin, after the Senate hearing in 2013, was now coming to a close. Aside from the reputational risks of Bitcoin, the main hurdle that most banks came up against, internally, was concern about money laundering. Regulators expected banks to keep track of the source and destination of all transactions going in and out, to ensure that the banks were not doing business with terrorists and mobsters. This was generally not hard because banks around the world were forced to keep records on all accounts and all transactions. But banks had faced billions of dollars in fines in 2013 for not adequately monitoring transactions coming from countries like Iran that faced economic sanctions. Many bank compliance officers determined that it would be all but impossible to know where money flowing into Bitcoin companies was ending up. Customers at a Bitcoin exchange could convert their dollars into virtual currency and then transfer the virtual currency to an unmarked address.

Jamie Dimon, the chief executive of the nation's largest bank, JPMorgan Chase, had told CNBC in late January that he was extremely skeptical that Bitcoin would ever amount to anything real. Dimon said that once Bitcoin companies had to follow the same rules as banks, when it comes to money laundering and compliance, "that will probably be the end of them."

Barry Silbert knew Dimon personally. When he saw Dimon's comments about Bitcoin, he quickly e-mailed Dimon a link to the pro-Bitcoin essay that Marc Andreessen had written in the *New York Times*. A few days later, Dimon called Silbert. Dimon had clearly read Andreessen's essay and sympathized with the view that virtual currencies could provide some opportunity for people outside the United States who didn't have access to good banks.

But Dimon responded that the potential of Bitcoin was not going to be enough to convince government officials to allow a competing currency to exist. Dimon knew what it was like to work in an industry that came under government supervision. Once Bitcoin came under similar regulation, it would require all the same fees and rules that bothered people in the traditional financial system. He didn't dismiss Barry's arguments, though, and invited him to come in and present Bitcoin to some of JPMorgan Chase's executives.

Dimon's perspective was representative of a broader shift in the banking industry's mind-set since the financial crisis. Before the mortgage meltdown had nearly brought down the American economy, Wall Street had hired some of the best young minds in the world and tasked them with finding innovative ways to make money. When many of those clever innovations ended up contributing to the economic collapse, the banks that survived were made keenly aware of how financial experimentation could go awry. What's more, regulators put in place a raft of new rules that forced banks to think twice before taking unnecessary risks. Just as important, government officials were forcing banks to pay billions of dollars in fines for past infractions. Few banks paid as high a monetary price as JPMorgan.

By the time Dimon and Silbert talked, the most important characteristic of any new business for JPMorgan was not how much money it would make, but how it would sit with regulators. JPMorgan had gone further than most in pulling back from potentially risky activity. During 2013 it had stopped working with remittance companies, check cashers, and even student-loan providers, not because it had to, but because it didn't want the headache. Other banks were taking similar, if less aggressive, steps.

As the comments at Lawsky's hearing suggested, this was nearly the opposite of the attitude in Silicon Valley, which had not been implicated in the financial crisis. The tech industry was increasingly confident about its own ability to change the world, emboldened by the success of companies like Apple, Google, and Facebook. Some of the most popular tech companies were ones such as Airbnb and Uber that openly challenged cumbersome regulations like those imposed on hotels and taxis. In the financial networks that Bitcoin was hoping to challenge, tech investors like Fred Wilson saw just another set of regulations that could be disrupted to create a more efficient market. If anything, the financial industry seemed even more open to disruption because the incumbent businesses were so afraid of breaking the rules.

Wences, who had been working at the intersection of technology and finance for two decades, acknowledged that for most of his career the center of power and wealth in the United States, and perhaps even the world, had been the financial industry and, specifically, New York. But he was outspoken in his belief that this was about to change.

"It's likely that the next twenty or thirty years are going to be the same for Silicon Valley," he liked to say. "In no other area are we going to see the passing of the baton so clearly as with Bitcoin."

The only problem for the Silicon Valley disrupters was that they still relied on banks to hold the dollars they used to pay their employees—and, in the case of Bitcoin companies, the dollars they received from customers to pay for the virtual currency.

Wences Casares had always used JPMorgan Chase as the bank for his previous startups—he had maintained an allegiance to the bank after it had given his first startup an account back in the 1990s. Now, though, when Wences applied to JPMorgan to open an account for his new company, Xapo, he was, for the first time, turned down. He found another bank that initially opened a corporate account for Xapo, but then shut it down right before Wences received a $10 million check from his new investors, the

venture-capital firm Benchmark. Wences was in the unusual position of having an enormous check and no one willing to accept
it. He was eventually saved by Silicon Valley Bank, the same bank
that was holding money for Coinbase and the only bank showing
any willingness to work with Bitcoin companies.

In the long run, though, Wences assured everyone he knew that
the cautiousness of the banks would matter less and less. At an
event hosted by JPMorgan in the Valley, to discuss Bitcoin, Wences
was dismissive when the topic of Jamie Dimon came up:

"I think whatever Jamie does or doesn't do will be as relevant
as what the postmaster general did or didn't do about e-mail."

CHAPTER 29

●

February 2014

Mark Karpeles was spending many of his days in early 2014 in a space on the ground floor of the Tokyo office building that housed Mt. Gox. Mark was turning the space into what he called the Bitcoin Café, a real-world showcase for Bitcoin in Tokyo—with a register that would be powered by a point-of-sale system that Mark had been designing. Mark was spending his time working out the details of the café, down to the programmable LED lighting on the ceiling and the recipes for the pastries that would be served. The café was almost ready to open, with wine on the shelves and light blue Bitcoin Café mugs sitting next to the register.

As he puttered around the café, Mark did not look like a man responsible for a financial company that was in the throes of an existential crisis. For most of January, the price of a Bitcoin on Mt. Gox had been almost $100 higher than on any other exchange. This was a result of the continued difficulty that Mt. Gox was having in transferring withdrawals to customers outside Japan. Mark blamed this on American banks, which refused to accept wire transfers from his Japanese bank. For all the people with dollars stuck at Mt. Gox it seemed that the only way to get money out was by using the dollars trapped in the exchange to buy Bitcoins and then transferring the Bitcoins out of Mt. Gox. The pressure of all these people trying to buy Bitcoins on Mt. Gox, with no ability to go elsewhere, allowed sellers on Mt. Gox to charge higher and higher premiums for their coins.

Then, in late January and early February, something even more worrisome started happening that sent the price heading in the other direction. The customers earlier in January had complained about the difficulty of getting dollars out of Mt. Gox, but now a growing number of Mt. Gox customers reported that they had requested withdrawals of Bitcoin and never gotten the coins. A few days after the hearings in New York, Mark put up a formulaic statement on the Mt. Gox website acknowledging the problem: "Please rest assured that this is only affecting a limited number of users and transactions, and that we are working hard on resolving this problem as soon as possible."

The thirty or so Mt. Gox employees in the company's Tokyo offices knew little more than Mt. Gox customers about what was going wrong. When Mark wasn't working on the café, he was in his office, behind a locked door on the eighth floor, far from the second- and fourth-floor offices where most of his staff was located. There were visible signs that all the stress was wearing on Mark. He was not yet out of his twenties but gray hairs were visible in his big black mane and he was clearly gaining weight. People in the office heard that Mark's Japanese wife had taken his young son and gone to live with family members in Canada, but Mark said nothing about it. Mark rarely interacted with his employees and maintained the same grip on the company's essential accounts that he had back in 2011 when Roger Ver came to help after the first big crisis at the exchange. The alienation from the ordinary world, which had helped lead Mark to Bitcoin, also made him a terrible person to run a Bitcoin company.

The Mt. Gox employees were as surprised as the exchange's customers when Mark decided, on Friday, February 7, to shut off all withdrawals from Mt. Gox. The panic that this caused only got worse on Monday when Mark provided the first explanation of what was going wrong. In a statement, Mark explained that the exchange had run up against a flaw in the Bitcoin protocol. The flaw, known as transaction malleability, allowed devious users to alter the codes that identified transactions in a way that made it impossible to tell if a transaction had gone through. Users in the know could request a withdrawal, change the code, and then request the same withdrawal again. Mark, in his statement, said this was not just a problem for Mt. Gox, but an issue with the Bitcoin software, which should have been fixed earlier.

The statement immediately sent the price of Bitcoin plunging on every exchange around the world—a flaw in the Bitcoin protocol could jeopardize everything. And Mark was correct that transaction codes had been susceptible to alteration for some time. What he didn't mention was that all the other major Bitcoin companies had known about the issue for years and had designed around it, generally by not relying on the transaction code in question. Gavin Andresen, the chief scientist at the foundation that Mark had funded, quickly came out swinging against Mark and said that the issue was not a bug, but a quirk, which others had dealt with easily. Mark came under withering attack from nearly every developer working on the Bitcoin software.

"MtGox tried to blame their issues by throwing Bitcoin under a bus and I am glad there has been a public rebuttal showing up their incompetence," one programmer on the developer e-mail list wrote.

After Mark publicized the issue, transaction malleability did, in fact, become a point of attack on the Bitcoin network. Bitstamp, the largest exchange, shut off Bitcoin withdrawals one day after Mt. Gox's announcement. But Bitstamp emphasized that it had lost no money as a result of the issue and, after putting together a quick patch, it was back up by the end of the week. Other exchanges remained open throughout. Mt. Gox, on the other hand, remained closed, creating a growing fear that something bigger was wrong.

WHEN MARK KARPELES showed up for work on Friday morning, his umbrella barely protected him from the unfriendly wet snow falling from the sky. He was wearing a short-sleeved shirt that hugged his round body, and he carried a large frothy coffee drink. Almost all the other exchanges around the world had recovered from the transaction malleability scare, but Mt. Gox showed no signs of allowing customers to again withdraw money. Mark's entrance to the building was blocked by a young man who had flown to Tokyo from London two days earlier to try to get some answers. With a sign in one hand that said, "Mt Gox WHERE IS OUR MONEY," the protester, a mustachioed programmer named Kolin Burges, placed himself in Mark's way and said, "Please, can I have a chat with you?"

Mark first tried to dodge him, but then stopped reluctantly when the man said, "I came all the way from London to try and get my Bitcoins from you—to find out what's happened."

"We can't do anything right now," Mark said, looking both disdainful and scared. He started again toward the door when Kolin asked the key question: "Do you still have everyone's Bitcoins?"

"Can you let me get inside please," Mark said as he tried to pass Kolin, who was bobbing and weaving to get in his way. "I'm going to call the police," Mark threatened, before Kolin finally let him pass.

Upstairs in the Mt. Gox offices, the staff didn't know any more than Kolin did about what was going on. They were still operating the exchange, allowing people to buy and sell Bitcoins with whatever dollars were still in their Mt. Gox accounts and taking in new deposits from daring customers. The price of a Bitcoin on the exchange fell lower and lower as people doubted they would ever be able to get the coins out. On Friday, the price stood at $300, half what it was on Bitstamp. Some people, including Roger Ver, were convinced that Mt. Gox's problems were temporary and jumped at the chance to buy coins on the cheap.

Mark would later say that during this time he was spending his daylight hours at the office and his nights at his apartment, alone with his cat Tibanne, furiously working his way through hundreds of pieces of paper containing the private keys to Mt. Gox's Bitcoin wallets. He had driven around in his car and collected the papers from the three locations in Tokyo where he had stored them (he had kept the keys on paper so they would not be vulnerable to hackers). Once he was back in his apartment with the QR codes—essentially complex bar codes—he began scanning in the private keys one at a time, with his computer's webcam. A combination of fear and sickness slowly overtook him as each one of the wallets he scanned in showed up on his computer screen as empty.

It would be hard for others to verify Mark's narration of what happened during those days because he kept such tight control over all the exchange's accounts. And as time went on, fewer and fewer people believed anything Mark said. But even if he was telling the truth, it was not what he told his employees and customers when he came in to work on Monday morning, ten days after Mt. Gox shut off withdrawals. In a public statement on the Mt. Gox site on Monday, he said, "We have now implemented a solution that should enable withdrawals and mitigate any issues caused by transaction malleability."

On the narrow Tokyo street outside the office, Kolin Burges maintained his one-man protest. There were still few Japanese

people using Bitcoin, but Kolin did attract a few foreign supporters who showed up as the week went on without any sign of a resolution to the problem. Mark had two security guards advise the
staff on how to deal with intimidating encounters. Mark himself
started taking taxis to work and leased space in an office tower
with better security. On Friday, Tokyo police showed up to remove
the protesters.

A few hours after the police left, the Winklevoss twins landed
in London for a weekend appearance at Oxford University. When
they turned on their phones in the plane, they found a worrisome
e-mail from Mark's deputy, Gonzague, with whom they had dealt
in the past.

"I would like to talk to you urgently regarding the situation
with MtGox," he wrote. "Would you mind signing this NDA and
call me ASAP on my mobile phone."

Cameron Winklevoss replied that a nondisclosure agreement
could be tricky, but he was happy to talk. After being out in
London all day, Cameron finally managed to connect with Gonzague by Skype when he got back to his hotel Friday night.

Gonzague got right to the point and explained the staggering
extent of the problem: some 650,000 Bitcoins—essentially all the
company's customer holdings—were gone, along with 100,000
coins that belonged to the exchange.

Cameron was stunned. Doing the most basic math in his head,
he knew that Gonzague was talking about hundreds of millions of
dollars worth of Bitcoins.

"How is that possible?" was all Cameron could ask.

Gonzague said that someone had been stealing from the company's online, or hot, wallet by changing the transaction iden-
tifiers. When the hot wallet was empty, Mark had unwittingly
refilled it with coins from the cold, offline wallets. Gonzague

told Cameron that Mark had continued doing this over and over again, until all the offline wallets were empty. The whole thing had been going on for months, or even years, and Mark apparently never realized it until now.

The explanation struck Cameron as implausible, but it didn't seem worthwhile to argue now. The bigger question was what was going to happen next.

Gonzague sounded oddly upbeat. He explained that Mark had "burned himself" and was agreeing to step aside, making it possible to move the business to Singapore and reincorporate under new owners, with the twins being obvious candidates. Gonzague thought it would be possible to do this without telling anyone what had happened. If the exchange could get an infusion of coins the business could make up the missing money over time, from fees. If this wasn't done, Gonzague said ominously, it could set Bitcoin back years.

It didn't seem like a terribly attractive business proposition to Cameron, but he wanted to hear more—if only to understand how bad this was all going to be for his Bitcoin holdings. He asked Gonzague to send him some sort of concrete plan for what they had in mind.

The next day Gonzague sent the twins a twelve-page document, labeled "Crisis Strategy Draft." It had been put together for Mark and Gonzague by a small public relations firm run by some Americans living it Tokyo. It was clearly a draft document, with typos and inconsistencies, but it pulled no punches about what had happened:

The reality is that MtGox can go bankrupt at any moment, and certainly deserves to as a company. However, with Bitcoin/crypto just recently gaining acceptance in the public eye, the likely damage in public perception to this class of technology could put it back 5~10 years, and cause governments to react swiftly and harshly. At the risk of appearing hyperbolic, this could be the end of Bitcoin, at least for most of the public.

After reading through the document, and its four-part plan for closing Mt. Gox temporarily and reopening it under new owners, the twins still couldn't figure out what was being asked of them, other than putting a lot of money into a failing company.

"I understand the larger points you raise, but it is unclear to me what the exact plan of action here is," Cameron wrote back.

The twins were not the only people to whom Mark and Gonzague were looking for a lifeline. They also sent the Crisis Strategy Draft to Barry Silbert in New York, who had his Bitcoin Investment Trust up and running with tens of thousands of Bitcoins. Essentially everyone told the Mt. Gox team the same thing: there was nothing to do but admit the losses and declare bankruptcy. When Roger Ver met the Mt. Gox team at the Tokyo American Club on Monday morning, he told them that no one in the world had enough Bitcoins to bail them out, except perhaps Satoshi Nakamoto. Mark and Gonzague didn't believe it, and wanted to keep the information in a small circle of people to give them more time to find a savior. After Mark refused to admit the problem in a call with members of the Bitcoin Foundation, Roger angrily called some of the foundation members himself and let them know what was happening.

Once the word spread among the top Bitcoin companies on Monday, they all began preparing for something that had the potential to take down the whole Bitcoin experiment. In a shared Google document, they worked on a joint statement that gave their best argument for why people should not lose hope. Ordinary

Bitcoin users got some indication that something was wrong when Mt. Gox's Twitter account suddenly disappeared on Monday. But Gonzague and Mark continued to hold out hope that someone would come in and bail them out. When Cameron wrote on Monday to ask what was going on, Mark said he was planning to begin talking with a bankruptcy judge on Tuesday. But, he emphasized, "Our current goal is to try to save MtGox before filing for bankruptcy—in which case filing wouldn't be required anymore."

The growing bubble of uncertainty over how this would all play out finally burst on Monday night when a popular Bitcoin blogger, known as the Two Bit Idiot, posted a leaked copy of the Crisis Strategy Draft. As it began to circulate and the Bitcoin masses tried to determine if it was legitimate, there was a sense of suspended motion on the forums and message boards, with everyone waiting for the bottom to fall out. The companies putting together the joint statement—Coinbase, Blockchain.info, BTC China, Bitstamp, and Jesse Powell's exchange, Kraken—were caught off guard by the leak and rushed to complete their statement, which ultimately came out a few hours after the leak. The companies urged Bitcoin owners to understand that the losses were the result of irresponsibility and bad behavior, not of a deeper flaw:

"This tragic violation of the trust of users of Mt. Gox was the result of one company's actions and does not reflect the resilience or value of Bitcoin and the digital currency industry."

The price did begin dropping on Bitstamp and other exchanges. But the free fall unexpectedly slowed within a few hours, before the price hit the low it had reached back in December when the Chinese exchanges turned off deposits. Many people seemed willing to believe the idea that there was nothing wrong with Bitcoin; there was talk that the disappearance of the most disastrous company ever to touch Bitcoin could end up being a good thing for the technology. If nothing else, people had invested enough time and money that they couldn't stomach selling out of a trough. By Wednesday morning, the price was back up where it had been when the Mt. Gox news came out.

Still, under the apparently calm surface, there was immense and largely unseen damage. As the enormous figures from Mt. Gox suggested, tens of thousands of people had kept their money with the exchange despite all the warnings, and those holdings, estimated at over $400 million the week before, had now disappeared in a mysterious puff of smoke. Roger had a Japanese friend, whom he had convinced to buy Bitcoins and who had left $12 million worth of coins with the exchange. The older man in Argentina who had purchased large numbers of coins from Wences Casares, back in 2012, had also kept them with Mt. Gox. The man had been using Bitcoin to keep his retirement savings out of the unreliable peso—but now it was Bitcoin that failed him. The man wrote in an e-mail to one of Wences's friends in Argentina that his life had been turned upside down by the event:

> I'll tell you that the collapse of Mt. Gox, where I had put absolutely all of my savings, left me more than demoralized. Not only because of the money, which was a lot, but because it destroyed the hopes I had created for using it as my wife and I got older. Each time this comes up it really hurts my health.

The same week as the collapse, lawyers in Chicago and Denver filed a lawsuit seeking class-action status to represent all the victims, and federal prosecutors were sending out subpoenas to aid in the criminal investigation they launched.

Even many of the victims blamed Mt. Gox rather than Bitcoin. Nothing had gone wrong with the Bitcoin protocol. In fact, Mt. Gox had long been held up as an example of the dangers that arose when Bitcoin users relied on central institutions, rather than the system of private keys and personal wallets that Satoshi had designed.

And yet, Bitcoin's standing as a universal money, answerable to no government—and beyond the reach of any one government—had opened the way for companies like Mt. Gox, companies that took advantage of the fact that in the Bitcoin industry, each person could make up his own rules. This wasn't a problem with the protocol but it was an issue with one of the central ideas that had motivated Bitcoin: the supposed benefit of releasing money from all the outdated rules and regulations that governed the existing financial system. Mt. Gox was, of course, not the first example of the dangers that arise in a system in which no one is responsible for providing oversight. An academic study in 2013 had found that 45 percent of the Bitcoin exchanges that had taken money had gone under, several taking the money of their customers with them. One of the most trenchant critics of Bitcoin, the *Financial Times* writer Izabella Kaminska, put it well in the days after the collapse:

> The only way to stabilise the system is to rid it of the "cheating incentive"—that being the incentive that encourages the "prisoner" to take the high-risk selfish strategy. Most of the time that depends on establishing a system of enforced protocols or regulations that penalise rulebreakers above and beyond the potential benefit of cheating.

Some of the recent converts to Bitcoin were not opposed to some sort of government oversight for this fledgling market. Ben Lawsky in New York used the incident to push ahead faster with his BitLicense. But it was somewhat unclear whether there would be anything left to license.

CHAPTER 30

●

March 6, 2014

I t was early in the morning, but a scrum of reporters had already gathered outside an unassuming three-bedroom house in Temple City, one of the many featureless towns that sprawled along the inland freeways heading east from Los Angeles, serving as magnets for upwardly bound Asian immigrants.

The reporters were chasing a story that would provide the Bitcoin world with a break from all the hard questions it had been facing. That morning, *Newsweek* had posted its first issue under new owners. On the cover was a dramatic mask, against a black background with the title "BITCOIN'S FACE: THE MYSTERY MAN BEHIND THE CRYPTO-CURRENCY."

Satoshi Nakamoto's identity had been a recurring fascination for journalists, but all the previous searches had ended with inconclusive results. Given Satoshi's skill in using anonymizing software, many assumed that Satoshi would never be found until he, she, or they decided to come forward.

The *Newsweek* reporter, Leah McGrath Goodman, had seemingly cracked the nut in the most unexpected way. The man she found was named Dorian Nakamoto, but the papers recording his immigration from Japan to the United States in 1959, at age ten, showed that his name, at birth, had been Satoshi. This Satoshi Nakamoto had gotten a degree in physics from California State Polytechnic University and had worked on classified engineering projects before his retirement. He lived with his mother and liked model trains, but his oldest daughter told Goodman that her father was a libertarian; his brother said Dorian loved his privacy. Dorian Nakamoto generally refused to speak with Goodman during the course of her reporting. But when she briefly confronted Nakamoto in front of his house to ask him about Bitcoin, he seemed to confirm the circumstantial evidence.

"I am no longer involved in that and I cannot discuss it," Goodman reported that Nakamoto told her. "It's been turned over to other people. They are in charge of it now. I no longer have any connection."

It was a completely unexpected outcome to the hunt for Satoshi—so unexpected that it almost seemed to make sense. A master of encryption would have used the most misleading disguise of all, hiding in plain sight with a number in the phone book. When some of the early Bitcoin developers who had corresponded with Satoshi talked with journalists that morning, they acknowledged that the story seemed to fit together.

"It's probably the best theory yet," Mike Hearn, the Google programmer in Switzerland, told one reporter.

When Nakamoto refused to come out of the house for much of the morning—despite being at home—it only seemed to confirm that he wasn't going to refute the story. For Hearn and many other Bitcoiners this was a terribly sad outcome. Satoshi had valued his privacy above all else and now that had been violated. *Newsweek* had even posted photos of the car in his driveway, with the license plates visible. It was particularly worrying because previous research had suggested that during the first year Satoshi had stock-piled Bitcoins that would now be worth nearly $1 billion, holdings that would make Nakamoto a target of any enterprising criminal. The death threats from fans of Satoshi started flowing into Good-man's inbox.

Eventually Nakamoto emerged from his house, and before he could shut the door, a crowd of reporters on his front porch clam-ored to ask him questions.

"Why did you create Bitcoin, sir?" one reporter shouted.

"OK, no questions right now," Nakamoto said, with a Japa-nese accent.

Nakamoto didn't want to talk; he wanted someone to take him to lunch. When someone else stuck a recorder in his face, he said: "Wait a minute, I want free lunch first. I'm going to go with this guy," pointing at a Japanese reporter for the Associated Press.

As he battled his way out onto the sidewalk, Nakamoto tried to shield his sleepy-looking eyes, behind big square glasses, from the sun. His floppy hair and loose-fitting pants and jacket suggested that he might not have spent much time outside. Looking for the reporter who had promised him lunch and clearly confused, he finally answered the question everyone was asking: "I'm not in Bitcoin—I don't know anything about it."

This was, as many reporters quickly pointed out, far from definitive proof that *Newsweek* had gotten the wrong guy. It is what many people assumed Satoshi would say if asked about his involvement in Bitcoin. Before the reporters could get more out of Nakamoto, he disappeared into the AP reporter's Toyota Prius and drove off toward a sushi restaurant. The other reporters jumped into their own cars and followed behind, rushing into Mako Sushi after Nakamoto. As the reporters barraged him with more ques-tions, he and the AP reporter left before ordering and returned to the car. What came next immediately entered the list of great

Los Angeles car chases, this one narrated in real time on Twitter by *Los Angeles Times* editor Joe Bel Bruno:

> There is a huge chase going on behind #Nakamoto. Tons of media. All heading west on the 10 freeway

> We think #Nakamoto might be heading toward downtown LA. Great American #Bitcoin Chase

> Traffic!!! Oh no #Nakamoto!

> We are two cars behind #Nakamoto, and it looks like the @AP reporter is doing all the talking. #Bitcoin

> Hang on folks. . . . .There might be some resolution here with #Nakamoto in downtown LA. #Bitcoinchase surrealer and surrealer

> So the Great #Bitcoinchase seems to have found a destination at the @AP bureau.

> But the #Nakamoto story isn't over. Hordes of media here waiting for him.

The reporters who had been part of the chase quickly parked and raced into the AP building. A few managed to squeeze onto the elevator with Nakamoto and the AP reporter. The reporters once again asked Nakamoto if he was the creator of Bitcoin and he once again denied it before disappearing into the AP offices.

With the reporters stationed outside the AP office waiting for Nakamoto's next move, the focus turned back to Goodman's article, which was now being looked at with a more skeptical eye.

Commentators on Reddit and Twitter pointed out that Goodman's evidence was almost entirely circumstantial, other than the quote she got from him in his driveway. As Gavin Andresen wrote on Reddit, in an angry open letter to Goodman, what she reported Nakamoto saying could "simply be an old man saying ANYTHING to get you to go away and leave him alone."

Several people were also combing through examples of Dorian Nakamoto's writing that had been found online. While the Bitcoin creator's early writing had been crisp and even elegant, Dorian Nakamoto's reviews on Amazon and his letters to a model-train magazine suggested a man with a mediocre handle on the English language. In an Amazon review of Danish butter cookies, he wrote:

> it has lots of buttery taste
> the shipment went well. i've had a nice comment from
> my kids. it's a perfect xmas and i would say, for
> other occasions.

As the afternoon went on, a growing number of people concluded that Goodman's article was aggressive journalism gone terribly wrong. The AP's story and video from its interview with Dorian Nakamoto did nothing to improve Goodman's case. Dorian clearly and explicitly denied that he had anything to do with Bitcoin. He seemed to have little familiarity with the technology, calling it "Bitcom" at several points, and implying it was a company at another point. The final piece of bad news for Goodman came that night, on the P2P Foundation website, where the creator of Bitcoin had posted a few items about Bitcoin back in 2009. In the first post since 2009—and the first communication from Satoshi in any form since 2011—the user Satoshi Nakamoto wrote five words: "I am not Dorian Nakamoto."

None of this evidence, in fact, proved that Dorian was not Nakamoto. If Dorian was Satoshi, he could have gone home from the

AP office, logged into his P2P account, and made the post. And if Satoshi was as smart as some people believed, he would have known exactly what to say to convince people he wasn't Satoshi (he would have also had to be a very good actor). But in either case, the events of the day underscored just how committed Satoshi still was to remaining anonymous. The reexamination of the evidence also pointed back to the hoard of Bitcoins that Satoshi had mined during the first year of the network's existence, when his computers kept the system running. An Argentinian security expert, Sergio Lerner, had done a thorough study tracing the patterns of Satoshi's mining during that time and concluded that he had captured well over a million Bitcoins, worth nearly $1 billion now. More impressive than that, though, was the security expert's conclusion, from a careful analysis of the blockchain, that Satoshi had never spent a single one of the Bitcoins he had created. His work in creating the system really did seem to be a selfless act.

In addition to what the day had revealed about Satoshi Nakamoto, the incident suggested that the identity of Satoshi Nakamoto really didn't matter much. For a few hours on the morning of March 6, the world had believed that the creator of Bitcoin was an aging libertarian and model-train enthusiast living with his mother. The price of Bitcoin didn't move much in either direction. The Bitcoin protocol was now maintained by Gavin Andresen and a team of developers and the code spoke for itself. Even if Satoshi had returned, it seemed he wouldn't have much to do.

FOR THE FUTURE of Satoshi's creation, the more important event on March 6 was one that few people knew took place. Just hours after the *Newsweek* headline started making its way around the Internet, four men took the stage at an auditorium in the New York headquarters of the Wall Street giant Goldman Sachs.

This was a private conference for some of the bank's most powerful hedge fund clients. In addition to appearances from former New York City mayor Michael Bloomberg, the former head of the Bank of England, and the former president of the World Bank, Goldman had put together a four-person panel to educate its clients on virtual currencies. The panel was led by the cohead of technology at Goldman, a tall, bald physics PhD named Paul Walker. He opened the fireside chat by describing the two things about Bitcoin that everyone seemed to be able to agree on: "It's something on the internet that seems to be worth money, and it seems to have been invented by a mysterious person." But, Walker said, in a joking reference to the morning's story from *Newsweek*, "the last part may no longer be true."

Sitting next to Walker were Barry Silbert and Chris Larsen. Larsen was the man Jed McCaleb had brought on to run his new cryptocurrency startup, Ripple. Most men in the room were wearing ties, but in true Silicon Valley style, Larsen and Silbert were not. The fourth member of the panel was the former head of the Financial Crimes Enforcement Network, or FinCen, James Freis.

Barry asked how many people in the room were skeptical about virtual currencies, and a good majority of them put their hands up. Barry noted how different this gathering was from the elite circles on the West Coast, where at recent events he'd attended a minority of the participants had expressed skepticism. Barry said it reminded him of the early days of the Internet when everyone in the tech industry was leaving good jobs to try to cash in on the new idea.

"It's either going to change everything, or nothing," Silbert said.

To appeal to all the financial minds in the room, Larsen said that all the early problems surrounding Bitcoin had obscured the fact that the technology underlying it made something possible that had never been possible before.

"The world now knows how to confirm financial transactions without a central operator," he said.

It was, though, Walker, the high-ranking Goldman executive, who provided the most encouraging comments about the technology. He said the conceptual advances made by Bitcoin weren't just clever; they were useful in ways that could influence the future financial system. He had obviously been spending a lot of time studying this and was clearly impressed by what he saw. He suggested that Goldman was not planning to buy or sell Bitcoins, but he indicated that the bank was taking a hard look at how the blockchain might be used to change basic things about how banks do business. It currently took the bank three or so days to settle stock trades. What if that could happen instantly and be recorded on a blockchain for everyone to see?

Barry Silbert and Chris Larsen were beaming. Few things could help a financial cause more than getting the imprimatur of the firm known as "the smartest on Wall Street," a bank renowned for always seeing what was coming around the next corner and making the right bets. Walker wasn't making any official announcements, but everyone could see the Goldman executive was into this.

Walker reflected an increasingly widespread fascination in financial circles with the blockchain concept underlying the Bitcoin technology. Many bankers had begun to understand what Gavin Andresen had seen back in 2010 when he first became entranced by the idea of a financial network with no single point of failure. For banks that were terrified of cyber attacks, the idea of a payment network that could keep running even if one player, or one set of servers, got taken out was incredibly attractive. More broadly, the banks were waking up to several increasingly viable efforts to decentralize finance and take business that had belonged to the big banks. Crowdfunding companies like Kickstarter, and peer-to-peer lending services like Lending Club, were trying to directly connect borrowers and savers, so that a bank was not necessary. The blockchain seemed to present a decentralized alternative to an even more basic part of the banking industry's business—payments.

The banks were notably not becoming any more friendly toward working with Bitcoin the currency. JPMorgan's operating committee, led by Jamie Dimon, decided in the spring of 2014 that it would not work with any Bitcoin companies. At events in California with tech moguls, Dimon spoke derisively about Bitcoin and the ambitions of Silicon Valley to take over Wall Street's business. Dimon said that JPMorgan and the other banks weren't going to go down without a fight. At one point, JPMorgan threatened to stop providing services even to other banks that had Bitcoin companies as customers—like the European bank working with Bitstamp. Other American banks went so far as to close down the accounts of individuals who transferred money to Bitcoin exchanges.

But inside almost all these banks, there were people who loved the concept of a decentralized financial system like Bitcoin. JPMorgan maintained a so-called Bitcoin Working Group, with about two dozen members from across the bank and around the world, which was led by the bank's head of strategy and which was looking at how the ideas behind Bitcoin might be harnessed by the financial industry.

This JPMorgan group began secretly working with the other major banks in the country, all of which are part of an organization known as The Clearing House, on a bold experimental effort to create a new blockchain that would be jointly run by the computers of the largest banks and serve as the backbone for a new, instant payment system that might replace Visa, MasterCard, and wire transfers. Such a blockchain would not need to rely on the anonymous miners powering the Bitcoin blockchain. But it could ensure there would no longer be a single point of failure in the payment network. If Visa's systems came under attack, all the stores using Visa were screwed. But if one bank maintaining a blockchain came under attack, all the other banks could keep the blockchain going.

For many technology experts at banks, the most valuable potential use of the blockchain was not small payments but very large ones, which are responsible for the vast majority of the money moving between banks each day. In the stock trading business, for example, the lengthy settlement and clearing process means that the money and shares are all but frozen for three days. Given the sums involved, even the few days that the money is in transit carry significant costs and risks. As a result, various banks began looking at ways they could use the blockchain technology to make these sorts of large transfers quickly and securely. For many banks, the biggest stumbling block was the inherent unreliability of the Bitcoin blockchain, which is, of course, powered by thousands of unvetted computers around the world, all of which could stop supporting the blockchain at any moment. This increased the desire to find a way to create blockchains independent of Bitcoin. The Federal Reserve had its own internal teams looking at how to harness the blockchain technology and potentially even Bitcoin itself.

Many in the existing Bitcoin community scoffed at the idea that the blockchain concept could be separated from the currency. As they viewed it, the currency, and the mining of the currency, was what gave users the incentive to join and power the blockchain. Given that a blockchain could be taken over and subverted if an attacker controlled more than 50 percent of the computing power on the network, a blockchain was only as secure as the amount of computing power hooked into the network. A blockchain run by a few dozen banks would be much easier to overwhelm than the Bitcoin network, which now commanded more raw computing power than all the major supercomputers combined.

Bitcoin mining, which had once been a thing that Martti Malmi and Gavin Andresen could participate in with just their laptops, was indeed well on the road to becoming an industrial enterprise. One of the big players was 21e6, the secretive project founded by Balaji Srinivasan and funded, in part, by Andreessen Horowitz. Balaji had been among the first to see that as the chips became more high powered, the factor determining who would profit from Bitcoin mining would be the energy costs involved in powering and cooling the chips. A chip that was fast but ate up energy and got hot—requiring cooling—could end up costing more in electricity bills than it earned in Bitcoins. To cut down on power costs, Balaji's team had designed a system that kept the chips immersed in mineral oil, which absorbed the heat and eliminated cooling costs. The data centers running 21e6 machines were now the single biggest source of mining power in the United States. And 21e6 was already working on its next generation of chips, with code names like Yoda and Gandalf.

In China some entrepreneurial young men with access to cheap hardware straight from the factories realized that their country provided its own advantage for cutting down on power costs: corruption. One mining operation near Beijing set up right next to a coal power plant, where it got its power practically free thanks to the relationship between the power company and the owner of the mining computers. Another so-called mining farm was set up in Inner Mongolia where cheap power was plentiful. Mining was particularly popular in China because it provided a way for Chinese citizens to acquire Bitcoins without going through the increasingly restricted Bitcoin exchanges.

Surpassing all these other mining operations, though, was a company created by a reclusive Ukrainian programmer, Val

Nebesny, who had designed several generations of ASIC chips after reportedly teaching himself chip architecture from a textbook. Initially, Val Nebesny and his business partner Val Vavilov had packaged the chips in computers that they sold to other Bitcoiners, under the brand name Bitfury. But over time the two Vals kept more and more of the computers for themselves and put them in data centers spread around the world, in places that offered cheap energy, including the Republic of Georgia and Iceland. These operations were literally minting money. Val Nebesny was so valuable that Bitfury did not disclose where he lived, though he was rumored to have moved from Ukraine to Spain. And Bitfury was so good that it soon threatened to represent more than 50 percent of the total mining power in the world; this would give it commanding power over the functioning of the network. The company managed to assuage concerns, somewhat, only when it promised never to go above 40 percent of the mining power online at any time. Bitfury, of course, had an interest in doing this because if people lost faith in the network, the Bitcoins being mined by the company would become worthless.

THE TWO VALS running Bitfury were rare as outsiders who were succeeding in the new, more sophisticated, and heavily scrutinized Bitcoin world. The Vals were certainly not entirely alone. Roger Ver, who had recently managed to renounce his United States citizenship, after years of trying, owned Blockchain.info, which was doing better than ever. The number of wallets hosted by the company had passed 1 million in January and in March was approaching 1.5 million. It became increasingly clear that Blockchain.info's careful structure—holding only encrypted files for its customers—allowed it to totally avoid the regulations coming down on other Bitcoin companies. Roger was constantly getting entreaties from venture capitalists who wanted to pay millions for some of his 80 percent stake in the company. Newcomers to the Bitcoin world were trying to emulate the Blockchain.info model and create technology that could allow Bitcoin to work as originally intended and escape regulations.

But most of the outsiders who had been pioneers in the early days of Bitcoin had not been able to transition to the new world. Charlie Shrem was sitting at home, under house arrest, while Mark Karpeles was dealing with prosecutors who were looking to punish him for the role he played in the ruin of Mt. Gox.

The early Bitcoin aficionados had certainly not gone away or lost heart. The online forums were still as lively as ever. But whereas these people had mixed and mingled with the big investors at the Bitcoin conference in 2013, they were now part of an isolated community that was cut off from the more sophisticated investors and programmers. This was not dissimilar to other protest movements that had sprung up after the financial crisis. Occupy Wall Street, which initially drew lots of attention—and raised issues that became a part of the national debate—ultimately splintered into many groups and disappeared from the public spotlight.

The marginalization facing the early Bitcoin community was on display at a conference for the more ideologically minded Bitcoin community in early March 2014, held by the Texas Bitcoin Association at a Formula One racetrack on the outskirts of Austin, Texas. Austin was a fitting place for the event because this was where Ross Ulbricht had grown up and founded Silk Road, the truest experiment in many of the early ideals.

Ross was now in jail in Brooklyn, awaiting trial, and his parents had moved to New York to be closer to him. But his mother, Lyn, returned to Austin for the conference. Now raising funds for Ross's legal defense, she explained that the Bitcoins Ross had when he was arrested had all been confiscated and the family was using its savings to pay for his expensive lawyers.

At the conference, she looked shrunken, but she was treated like an honored guest and she delivered greetings from Ross, who called her frequently from prison in Brooklyn, where he said he was doing well, practicing yoga, and serving as a tutor for other inmates as he awaited trial. The market that Ross had created was generally viewed, in this crowd, as a moral good that had allowed people to make their own choices about how they wanted to live, without government intrusion. Rather than doing any sort of evil, Silk Road had made the world a safer place by allowing people to buy their drugs from the safety of their home.

The accusations that Ross had solicited assassins to murder people were more divisive. In legal papers, prosecutors in Maryland charged Ross with hiring the Silk Road user nob (actually an undercover agent) to murder Curtis Green. Prosecutors in New York accused Ross of hiring the Silk Road user redandwhite to kill several Silk Road scammers. But there was no evidence in either case that anyone ended up dead (in the redandwhite case, Canadian police could not turn up anyone matching the names of the people Ross allegedly tried to have killed). What's more, in the indictment moving toward trial, these accusations of murder for hire were not included as formal charges. Ross's mother said it was terribly unfair for prosecutors to pin these accusations on Ross if they were not willing to charge him. But even if Ross had done the things the agents claimed, there were plenty of conference attendees willing to argue that he had made the right decision.

"What if the scammer was going to expose every Silk Road customer?" one young man asked at one of the conference happy hours. "He was doing no one any good. Ross did something to protect all of those thousands of customers."

Aside from Lyn Ulbricht's appearance, the most memorable part of the conference was Charlie Shrem's virtual appearance. He couldn't travel to Texas, of course, but the organizers got him on

Skype and projected a live feed of Charlie, from his basement bedroom with his guitars behind him. Charlie was wearing a brown "BOUGHT WITH BITCOINS" T-shirt that he'd worn two months earlier when he met Nic Cary for drinks, before his arrest.

Charlie was in the midst of trying to negotiate a settlement with the government to lessen the time he'd have to serve—eventually, as a result of these talks, Charlie would plead guilty to one count of aiding and abetting an unlicensed money-transmitting business and accept a one-year prison sentence. In the meantime, his lawyer had told him to avoid making any public statements that might hurt the talks. But his loquaciousness and desire for attention were irrepressible. This kid, who had once been called Statist for his mainstream politics, now gave a fiery talk that was a play on a well-traveled speech delivered by the founder of the Pirate Party several years earlier.

Friends, citizens, Bitcoiners, there is nothing new under the sun.

My name is Charlie Shrem, and I speak to you from under house arrest.

During the last few weeks, we've seen several examples of legal outbursts. We've seen the police abusing the measures available to them. We've seen the actions of the financial services industry. We've seen high-profile politicians mobilizing in order to protect the financial and banking industry.

All of this is scandalous without parallel. That is why I stand here today.

When it ended, some twenty minutes later, there was a smattering of applause and shouts. Charlie complained that his connection was making it hard for him to hear the crowd's response. But the people who got up to ask him questions told him he was a hero.

"We all love you. You are still a huge part of this community," said the shaggy-bearded founder of a Bitcoin charity. "What kind of beer should we send to you? Because you said you were looking for six packs."

"I love Blue Moon, but anything exotic is good," Charlie said.

"All right, cool. Stay strong Charlie!" the man shouted with a raised fist.

ALMOST NONE OF the more recent, moneyed arrivals at Bitcoin showed up for the conference at the race track. But many of them did fly to Austin just as the conference was ending, to attend another conference, SXSW, the storied public gathering where Silicon Valley mingled with celebrities. On the first day of SXSW, in a marquee session with Google chairman Eric Schmidt, Google's "director of ideas," Jared Cohen, responded to a question about Bitcoin with his conclusion: "I think it's very obvious to all of us that cryptocurrencies are inevitable."

Fred Ehrsam, the former Goldman Sachs trader who had joined the Bitcoin company Coinbase a year earlier, was given the honor of his own SXSW session—not a shared panel with other entrepreneurs—and it was put in one of the largest rooms in the convention center, which quickly filled up. In the question-and-answer session that followed Ehrsam's talk, Lyn Ulbricht was the first one in line at the microphone. She said something about using Bitcoin for charity, but she was clearly there to make a plug for Ross's legal defense fund, which she told Ehrsam was hosted on Coinbase. Whereas Lyn had been a star at the Bitcoin conference, here she was an unhappy reminder of a side of Bitcoin that Ehrsam and others wanted to put behind them. Fred responded politely and fumbled to find something to say about the value of Bitcoin for charitable donations broadly. But Fred was not shy about his belief in the transformative impact that Bitcoin would ultimately make as it became "the prevalent transaction medium on the internet."

Fred's biggest backer, Marc Andreessen, was increasingly vocal about his belief that the Silk Roads of the world were quickly giving way to more Coinbases. Andreessen frequently noted that in the early days of the Internet, when he was creating the first web browser, the new technology had lacked the infrastructure that would have made it appealing to a mainstream audience, and so it was relegated to fringe groups that were willing to experiment with new technology. In time, though, "the fringe characters tend to get alienated and then tend to move on to the next fringe technology."

"You don't get the new technology from the mainstream," he said. "My prediction is actually that the libertarians are going to turn on Bitcoin. I think that's about two years out."

SXSW underscored how thoroughly Silicon Valley was winning the battle to shape and define the Bitcoin technology. The gathering also served as a stark reminder of how Silicon Valley had, more broadly, emerged as the big winner after the financial crisis. With Wall Street in retreat, these were the new billionaire power brokers, flying around the country in private jets. On Saturday night Ehrsam was invited to an exclusive party hosted by Andreessen Horowitz. At the party, which was attended by celebrities like Ashton Kutscher, Ehrsam talked about Bitcoin with Ben Horowitz and the rapper, Nas, whom Horowitz had brought on as an investor in Coinbase. The big names like Horowitz at SXSW reiterated what world-changing new technologies, such as Bitcoin, the tech industry was helping to bring to the world. In an onstage conversation between Nas and Horowitz, Horowitz called Bitcoin "the internet of money," with the potential to help billions of people. Andreessen Horowitz had recently closed a $1.5 billion fund, and the partners said privately that they wanted to spend as much as $200 million of that on Bitcoin and blockchain startups, if they could find deserving ones.

But the week in Austin couldn't help fueling suspicion that perhaps, as in the old way of doing things, the economic benefits of all the new technology were, at least so far, accruing to only a small elite, while the 99 percent that Occupy Wall Street had worried about were left reading about it at home on Reddit and Twitter. Bitcoin itself faced the same concerns. Years earlier, Bitcoin had promised that it would spread its benefits to all its users, but by 2014 large chunks of the Bitcoin economy were owned by a few people who had been wealthy enough before Bitcoin came along to invest in this new system. Most of the new coins being released each day were collected by a few large mining syndicates. If this was the new world, it didn't seem all that different from the old one—at least not yet.

CHAPTER 31

●

March 21, 2014

Many of the early adopters who had managed to stick around and make something of themselves flew out for the second occurrence of Bitcoin Pacifica at Dan Morehead's vacation home on Lake Tahoe, where a large staff catered to the crowd's every desire, allowing Morehead to play the relaxed host in his elegant black loafers and a pinkish red shirt that set off his perfect tan.

Among the guests was Jed McCaleb, the founder of Mt. Gox, who had recently been helping Morehead's firm look for new Bitcoin investments. Jed spent a lot of time at Morehead's house talking to Jesse Powell, someone he had first met at the 2011 Bitcoin conference in New York. Jesse, who was sporting sweatpants and athletic socks, was still working on the exchange that he had begun building after traveling to Tokyo in 2011 and seeing what a mess Mt. Gox was. Three of the young men who ran the successor to Mt. Gox, Bitstamp, had flown in from Slovenia and were buzzing about the matching Teslas they had recently purchased with some of the profits from their business.

Roger Ver couldn't make it to Tahoe. He had recently renounced his American citizenship and become a citizen of Saint Kitts-Nevis, which offers passports to people who buy at least $450,000 of real estate on the island. Roger had applied for a visa to come to Morehead's event, but the American government had denied the request. Roger's old friend Erik Voorhees was in Tahoe, up from Panama where he was spending his time dealing with the Securities and Exchange Commission investigation of the shares he had sold in SatoshiDice. Erik had come to be viewed as one of the few people who managed to remain ideologically engaged without letting ideology totally overwhelm their business instincts. The company that Erik founded after leaving Charlie Shrem's BitInstant, Coinapult, was aiming to make it easier to send Bitcoin by e-mail and text message. But the conflict between ideology and commerce had, in fact, become too much for Erik to bear. The investigation by the Securities and Exchange Commission had forced him to sell some of his Bitcoin holdings to pay for a lawyer. He worried that if he continued to speak out politically his company would become a target of government officials. Rather than drawing back from the politics, he had decided to leave his company and move with his fiancée back to Colorado.

"The way I felt I could contribute best is by being a very outspoken advocate for what Bitcoin stands for," he said.

For many of the attendees, though, the biggest celebrity at the gathering was a reclusive man who was essentially unknown to the outside world. Nick Szabo had been deeply involved with the Cypherpunks back in the early days and in 1998 had invented bit gold, one of the most commonly cited forerunners of Bitcoin. More recently he had become, for many Bitcoin insiders, the most likely candidate for Satoshi Nakamoto.

Nick was nearly as mysterious as Satoshi himself. He kept a blog where he occasionally wrote learned essays on topics like online security, monetary history, and property law. But there was no public record of where he worked and lived, and some people questioned whether he was a real person. Nick's writing, though, would put him on anyone's short list for Satoshi. Back in the 1990s, he wrote more than just about any other Cypherpunk about the promise of digital money, culminating in his proposal for bit gold. Just a few months before Bitcoin was released, in April 2008, Nick had posted on his blog an item in which he talked about creating a trial model of bit gold and asked if anyone wanted to help him "code one up." In August of that year, at the same time that Satoshi was privately e-mailing Adam Back about Bitcoin for the first time, Nick offered on his blog to sell some old collectible private banknotes, to help deal with "personal cash flow needs." At about the same time, he wrote a burst of blog posts about the history of money, smart contracts, and bit gold, and said that if he could make bit gold work it would be the "first online currency based on highly distributed trust and unforgeable costliness rather than trust in a single entity and traditional accounting controls."

When Satoshi's white paper came out publicly three months later, it cited two other obvious forerunners of Bitcoin—b-money and hashcash—but did not cite Nick's work. During this period, Nick maintained what many people later came to think was a rather suspicious silence, despite the fact that this was a project that he'd been involved in for over a decade. Most bizarrely, Nick altered the dates on his 2008 postings about bit gold to make it appear as though they had been published after Bitcoin was released, rather than before.

Not long before the Tahoe gathering, a blogger who went by the name Skye Grey had posted two persuasive essays comparing Nick's online writing with that of Satoshi, and concluded that the similarities in style and word choice were unlikely to be a coincidence. Both Nick and Satoshi, Skye Grey wrote, made "repeated use of 'of course' without isolating commas, contrary to convention" and "repeated use of 'timestamp' as a verb," among other such tics. Then there were smaller eyebrow-raising details, like Satoshi Nakamoto's initials being a transposition of Nick Szabo's.

Nick had made a brief statement, by e-mail, to deny that he was Satoshi, but that didn't quiet the speculation. At Morehead's gathering, people spoke in hushed tones about things they'd overheard Nick saying. Nick showed up at Morehead's private gathering because a few months earlier he had quietly joined a cryptocurrency startup that was operating in stealth mode. The startup, Vaurum, was based a few blocks from Wences's office in Palo Alto and focused on the task of matching up big holders of Bitcoin wanting to buy and sell. Nick, though, had joined Vaurum to do more sophisticated work on so-called smart contracts, which would allow people to record their ownership of a house or car into the blockchain, and transfer that ownership with the use of a private key, something Nick had been thinking about for over a decade. This was the kind of thing that Satoshi was writing about at the beginning, but Satoshi had believed that these more advanced uses of the blockchain would take off only after Bitcoin caught on as a currency.

At Morehead's house, it was obvious that Nick was a guy who lived a life of the mind. His large frame was covered haphazardly with old jeans and a flannel shirt. His beat-up black sneakers looked as if they'd been purchased back in the days of DigiCash. His hair was an unkempt ring around his scalp, not unlike a monk's tonsure just after a long nap.

In Tahoe, Szabo didn't seek out conversation and didn't make much eye contact when engaged. He had a seemingly perpetual smirk on his sleepy, bearded face. Most of the other attendees watched him from a distance, waiting for him to open up. During the cocktail hour before dinner, on Friday night, when the topic of Satoshi came up in the small group where he was standing, he took the opportunity to sound off on all the mischaracterizations of him, including the frequent descriptions of him as a law professor at George Washington University—and the notion that he created Bitcoin.

"Well, I will say this, in the hope of setting the record straight," he said with an acid note in his voice. "I'm not Satoshi, and I'm not a college professor. In fact I never was a college professor. How the media got a hold of that, I don't know."

"Even I thought you were a college professor," a New York trader, standing next to Nick, said with a laugh.

Nick did use a George Washington e-mail address, but he explained that this was because he had gone to law school at the university in mid-career, "just for the reality check of what I'd been thinking about." He had paid the tuition thanks to some stock options he had from his earlier days as a security programmer. He had returned to school in part because he had become convinced that the singular focus on markets, among libertarians and crypto-anarchists, was naive. Szabo believed that society had multiple "protocols" beneath markets, such as the legal system, which determined how markets worked. All of this, though, had just been a hobby for Nick, until very recently.

"The cryptocurrency economy is actually big enough that I can actually make a living out of it," Nick said with a bit of a chortle.

As he walked over to the big living room, for dinner, Nick explained that he traced the germ of all this back to his childhood in Washington State and his father, who came to the United States after fighting in the 1956 Hungarian revolution, which the Soviets crushed.

"We're fairly rebellious sorts," he said of his family. "To really have the freedom to be creative you have to think outside the box."

This was about as personal as Nick got in discussing his motivations. He was a person who liked thinking about the world—not himself—and this is one of the most useful characteristics for someone trying to create great things.

At dinner, everyone was too polite to speculate about Nick, but the *Newsweek* story of a few weeks earlier naturally kicked off conversations at the different tables about Bitcoin's origins.

"Is there no doubt in any of your minds that maybe this was a product of the NSA?" asked the New York trader who had been talking with Nick before dinner.

Erik Voorhees scoffed and said that the government would have been unlikely to come up with something so brilliant. But the trader cited his own work experience at the NSA, and said Erik was underestimating the level of intelligence the NSA attracts. Erik, always willing to listen and learn, said that if it was the NSA, "it is the best thing the government has ever done."

Erik's pet theory was that Satoshi was actually a small circle of programmers at some major tech firm, who had been assigned by their company to come up with a new form of online money. When the project had come back and was deemed too dangerous by the higher-ups the creators decided to put it out anonymously— they "felt really strongly that this was something important they discovered and went rogue with it," Erik explained, even while noting, with a laugh, that he had no actual evidence to back up his hypothesis.

Most of the weekend, though, was spent talking not about Satoshi, but instead about the incredible challenges that everyone in this group faced. The one-two punch of Charlie Shrem's arrest and Mt. Gox's collapse had killed much of the hope that Bitcoin would gain mainstream acceptance anytime soon.

Dan Morehead had been running his Bitcoin operations from inside Fortress's San Francisco offices, and there had been a vague plan for his small team to be integrated into Fortress, a publicly traded company. With all the crises, though, Pete Briger had let Dan know that Fortress was not going to be able to have a formal role. Dan was going to have to move his staff, operating under the name of his old hedge fund, Pantera Capital, out of Fortress's offices.

Things did seem to be going well for the old college fraternity brothers who founded Bitpay, both of whom were in Tahoe. They had signed up lots of new online merchants who were happy to find a cheaper way to process online transactions—the 1 percent that Bitpay charged versus the 2 to 3 percent charged by credit cards—without worrying about chargebacks. But it was now becoming evident that consumers had much less of a reason than merchants to use Bitcoin for online purchases. Consumers, after all, never see the 2.5 percent processing fee that merchants pay, so products aren't cheaper when purchased with Bitcoin. And consumers generally like having the peace of mind offered by chargebacks. For the sake of Bitcoin as a whole, there were many who worried that the consumers who were buying things online through Bitpay were pushing the price of Bitcoin down; generally when online retailers accepted Bitcoins they immediately sold them off for dollars, creating a downward pressure on the overall price.

Bobby Lee talked at Tahoe about the many unusual stresses of running a virtual-currency startup in China. After the government had forced the payment processors to cut off Bitcoin exchanges back in December, Bobby's competitors had quickly opened bank accounts where customers could deposit funds. Bobby had chosen not to follow the same path—it seemed to violate the clear intent of the statement from the Chinese regulators in December. Bobby had grown up working for American companies, which generally tried to obey, or at least give the appearance of obeying, not just the letter but also the spirit of the rules. Bobby had internalized this cultural code. But as Bobby watched his business dwindle, and his competitors thrive, his Chinese cofounders pushed him to understand that Chinese regulators weren't looking to enforce a strict reading of the law—they just didn't want to have anything shoved in their face.

"Turns out, in China, there's no ethics—there's no moral obligation," Bobby would say of his discovery, with a hint of amusement and a dash of frustration. "Westerners see that as a bad thing. Chinese see that as, 'We're being flexible.'"

With a sense that he was caught in a street fight and limiting himself to punching with boxing gloves, Bobby eventually bent to the Chinese way of doing business and opened up the company's bank accounts to customer deposits shortly before coming to Tahoe.

"If no one listens, and there is no penalty, our competitors do what's best for them and then we're left in the dust," Bobby explained. "So instead we decided to embrace the local method."

There were, though, limits to how far Bobby would go in his hunt for business. He was outspoken about his belief that his competitors were faking their volume numbers to make it look as though they were attracting more business. He also initially declined to follow the lead of one of his increasingly successful competitors, OKCoin, which had introduced what is known as margin trading. Customers of OKCoin could essentially borrow money to make bigger bets on Bitcoin. If the price went up, customers could pay back the borrowed money, but if it went down the customers quickly lost their original money—the normal outcome in margin trading. This didn't seem to Bobby like a good formula for a long-term business, though he was coming to reconsider all of his Western judgments.

Despite all the challenges, Bobby was clearly having a good time, enjoying the audacity and inventiveness that were required of an entrepreneur in China. He was making plans to move his staff into bigger offices and he had announced his candidacy for one of the Bitcoin Foundation seats that Charlie Shrem and Mark Karpeles had vacated—a seat he would eventually win. In Tahoe, he was the very picture of the fun-loving, confident risk taker, sweeping the poker games. He likened his situation in China to being in a tunnel with no clear way out.

"Everyone behind me is like, 'Dude, Bobby it's a dead end, you are not going to get out,'" he said. "But I'm like, 'If I get out, the prize is so huge."

The weekend provided plenty of reminders of why everyone had gotten into this in the first place. After dinner on Friday night, Dan introduced a celebrated economics professor at Stanford, Susan Athey, a winner of the most prestigious award for young economists, who had recently been diving into the blockchain technology. She told the group of her discovery of Bitcoin in the spring of 2013. At the time she went to her academic colleagues and found that "none of them could wrap their head around it." That provoked her to look more deeply, and as she did, she slowly came to understand the potentially enormous implications of the technology:

"We all hear the store of value. Here's a way to move money and to buy things outside the law. Maybe it's a competitor to fiat currency. Is it a disrupter to the traditional banking sector; an enabler of e-commerce and remittances; a superior internal ledger system for multinationals? That's not what all the reporters are asking about but that's another possibility that we see.

"By the time I felt like I really understood it I was really excited to share that knowledge, and discuss it with a wider audience," she said. "You want everyone to understand it too so that they'll really appreciate the really massiveness of this innovation.

"It's not just a thing, it's a phenomenon."

•

GAVIN ANDRESEN HAD been invited to Dan Morehead's house in Lake Tahoe, but he had elected to stay home in Amherst. He was receiving many invitations to swanky gatherings and turning essentially all of them down—though he did accept an invitation to speak to the local Rotary Club. When he had been asked to attend the prestigious Aspen Institute, a friend had urged him to go.

"It will change your life," the friend told him.

"I don't want my life to change," he responded. "I like my life."

He had certainly profited from Bitcoin's rise: he had been paid by the foundation in Bitcoins since 2012 when each Bitcoin was worth $10. His wife had pushed him to use some of the money to get his own office in downtown Amherst, and a second car for the family. But the car they chose was a modest black Nissan Leaf. And for an extravagant family vacation, he planned a trip to visit his mother in Washington State for a Women's Auxiliary ceremony. For the first time, Gavin hadn't worried about the prices of the hotels he was booking, and he planned a helicopter trip for his family to see Mount Hood.

Gavin was similarly understated about Bitcoin. He still lived for the project, but like other developers he was deeply aware of the flaws that still existed. He called the software that Satoshi had created a "hairball" containing lots of different things stuck together. As he saw it, the volunteer developers were still trying to untangle it. He was particularly focused on the limited number of transactions that were being confirmed and recorded on the blockchain with each new block. On average, there were only about four hundred transactions getting confirmed every ten minutes in mid-2014. If Bitcoin wanted to compete with payment networks like Visa, which processed two thousand transactions each second, the software was going to need to change significantly.

Among the broader community of Bitcoin programmers there was constant griping about the increasing centralization of the entire Bitcoin ecosystem. The network had been designed to encourage all of its users to participate. But now, only people with access to super-powered computer chips and cheap energy were able to take part in the mining and transaction recording process—something that a small handful of companies were dominating. As had happened with several previous decentralized systems, this one had naturally tended toward greater centralization because of the efficiency made possible by specialization. This looked, increasingly, like Napster giving way to iTunes. In that case, the old power brokers—the record labels—were destroyed, but they were mostly just replaced by a new set of power players.

Gavin rarely brought it up publicly, but there was another, more frightening problem that didn't appear to have any immediate solution. There were a growing number of examples of Bitcoin being used by criminals to demand and collect ransom, which was much easier with Bitcoin than with traditional means of payment. When criminals accepted cash for ransom they had to physically collect the money at some point, which provided some indication of their location. If ransom was sent digitally via PayPal, it didn't require a physical handoff, but the payment could later be reversed. With Bitcoin, criminals could demand that a victim send money remotely, and once it was sent, there was no reversing it. The previous fall, a malware program known as CryptoLocker had surfaced, which had the ability to seize computers and lock the hard drive until a Bitcoin ransom was paid. The fears about ransom were a large part of the reason that many Bitcoiners had been angry at *Newsweek* for "outing" Dorian Nakamoto. If he had really been Satoshi, his outing would have made all of his family members unusually vulnerable to kidnapping and demands for payoffs of various sorts.

Gavin didn't know it, but for months, a hacker demanding ransom was targeting Hal Finney and his family, despite the fact that Finney had been rendered almost entirely unable to move or communicate by his disease. The attack came to a terrifying climax when the hacker called the police and reported that a murder was taking place at Hal's house; this forced the local police and fire department to evacuate Hal and his family, a taxing experience that came just a few months before his death. Roger Ver had dealt with what appeared to be the same hacker, but beat him off after offering a public bounty for his capture. The best solution to this threat seemed to be wallets that were programmed to allow for reversible transactions. In the meantime, many Bitcoin developers emphasized, whenever possible, that they did not keep most of their money in Bitcoins.

The developers, though, appeared to have a staying power that eluded many of the other early adopters of Bitcoin, in large part because of their more practical approach to the project. Jeff Garzik, the programmer in North Carolina who had gotten involved back in 2010, had been hired by Bitpay to work on the Bitcoin protocol full-time. Martti Malmi had recently quit his job in Helsinki after a new payments startup invited him to come on board, knowing about his history with Bitcoin. Adam Back, the creator of hashcash back in 1997, had recently started working with an investor on a bold new project that aimed to make it possible to take Bitcoins off the main blockchain and on to so-called sidechains, where new applications could be built.

The small team of core developers working with Gavin was made up of people who had gotten involved back in 2010 or 2011 and managed to stay out of the spotlight almost entirely—men like Gregory Maxwell and Wladimir J. van der Laan. The person responsible for writing the majority of the updated Bitcoin core protocol was a thirty-year-old Belgian whom many Bitcoiners had never heard of, Pieter Wuille.

It came to seem that the people who wanted Bitcoin to do the least for them were the ones who were managing to do the most for Bitcoin.

WENCES CASARES WASN'T looking for Bitcoin to change his life, but he was still imagining that Bitcoin would change the world. His passion for the project had continued to win over important new supporters. Max Levchin, the cofounder of PayPal, and one of the skeptics back at the Allen & Co. conference in Arizona in 2013, had been brought around by Wences at the 2014 version of the conference and was now coming on board as an investor in Xapo. Wences also knew from his friend David Marcus that PayPal was moving toward integrating Bitcoin into all of its online products, making the virtual currency available to a much broader audience.

But the day-to-day work of moving his own Bitcoin company forward was going much more slowly than Wences had expected, largely because of the continued skepticism in the traditional financial world. In April, Wences announced that Xapo would be releasing the first Bitcoin debit card with MasterCard, but almost as soon as the announcement went out, MasterCard called and told Wences that the project had not been approved at the highest levels and was now being killed—a public relations snafu for Xapo. Wences himself was constantly flying around to appease the latest bank to decide that it was going to close down the accounts of Xapo or some other Bitcoin company that Wences was helping out.

In the midst of all this, in June, Wences took one of his periodic trips to visit Xapo's operations in Buenos Aires and the old friend who oversaw it all, Fede.

As on every trip home, Wences had to confront the frustrations of Argentina's broken financial system. This time around,
he wanted to buy a car so that he could travel to and from a property he'd recently purchased in Patagonia. As with most big-ticket
items in Argentina, the seller would accept only cash. Because
Wences still didn't have an Argentinian bank account, he had to
go to a specialized money changer who had a bank account in the
United States and could accept a transfer of dollars from Wences's
American bank account and pay out to Wences in wads of cash.
This served as yet another reminder of why he was working on
Bitcoin.

The scale of Wences's ambitions was evident inside the Xapo
offices, which were packed with young programmers. One was
working on a Hindi-language site, which would make Bitcoin
available to people in India, widely seen as one of the biggest potential markets given the Indians' levels of computer literacy and
the amount of remittances that were sent from Indians abroad.
Another programmer was building an application that would
allow people anywhere in the world to find people near them looking to buy or sell Bitcoins. At this point Xapo was still primarily
used by big institutional investors who wanted the best possible
security for their millions of dollars of coins. But the Xapo team
was trying to make the service more accessible to smaller holders,
and many people were eager for secure storage after the collapse
of Mt. Gox.

On one of the first mornings Wences was in Buenos Aires,
the team of programmers had a videoconference with the Xapo
staff in Palo Alto. The team in California had just moved to much
larger offices above a bank. These staffers now had a whole floor
to themselves, with windows wrapping around the entire office.
The Americans, who generally dealt with the business side of the
operation, rather than programming, ran through all the new

agreements they were working on. They were talking with AIG about insuring all the coins in the vault against losses, and with three different banks about taking deposits from customers.

"We're in a really good position in comparison to a lot of people in the industry in respect to banking relationships. Most people are just hoping to get one," one of the employees in California said.

They also were working with a debit card issuer in Gibraltar after the problem with MasterCard earlier in the spring, and were hopeful that they would be able to distribute the cards worldwide.

After lunch, Fede got the keys for the Buenos Aires staff's new, larger office, which was two flights down and occupied an entire floor, with big conference rooms and a Ping-Pong table. While the staff gleefully ran around the empty offices like schoolchildren, Wences sat down in the glass-enclosed conference room. He looked exhausted. He explained that he had expected some kind of respite once he sold off Lemon in the winter. But before he'd been able to come up for air, he was back under, trying to get Xapo running, and dealing with the unending series of crises that seemed to be an endemic issue for Bitcoin companies.

The problems, though, seemed to Wences only like more evidence of why Bitcoin was necessary. In the current system, financial institutions were given the power to determine what sorts of businesses could live and die. His vision for what Bitcoin could do had remained steady. While others were talking about micro-payments and smart contracts, he was still fixated on the idea of a digital gold that people anywhere in the world could hold without requiring any permission from anyone. This was still the kid who had grown up in Argentina, watching his family look for a place that was more secure and reliable than the peso to store their savings.

It might have just been the exhaustion, but Wences was sourly dismissive of all the talk about Bitcoin's potential as a new payment system. He was an investor in Bitpay but he said that fewer than one hundred thousand individuals had actually purchased anything using Bitpay.

"There is no payment volume," he scoffed. "It's a sideshow."

The real story, he said, was the steady viral growth that had already taken Bitcoin, by Wences's count, from a few people on that first day back in January 2009 to six million users.

"People buying half a Bitcoin, storing it, treasuring it, and talking about it—and getting more than one person in," he said. "That's all Bitcoin has been about for four years—and that's all we need to get to where we want it to be."

He did believe it would eventually be the best payment network the world had ever seen. But that would happen only when a billion people owned some Bitcoin. He made the familiar comparison to the Internet in 1993. Back then, he had crowed to his mother when he got one of the first ten million or so e-mail accounts, which allowed him to exchange messages with a professor in North Carolina. His mother had derided it as a curiosity: how would it help her communicate with anyone she knew? But Wences believed back then that the ability to freely send information to anyone, anywhere in the world, would eventually matter. And he ended up being right. Now he believed that the ability to send money to anyone, anywhere in the world, free would eventually matter.

"I thought I was lucky to have lived through that once—and I can't believe I get to see it again," he said. "This is just the spot. It feels exactly the same way—it was so hard to explain."

In the meantime, he said there would be setbacks as governments banned it and banks made it harder to transfer dollars and pesos to Bitcoin companies.

"I'm patient. This takes a decade, or two decades. I'm not going to go home because this takes one more decade."

From Buenos Aires, Wences flew to Brazil for his first vacation in what seemed like years. Belle and the three children met him and they stayed at a house near the beach in Rio and caught all the World Cup games they could. But even before the World Cup was over, Wences and the family were up in Utah for the latest exclusive conference held by Allen & Co., this one an even higher-profile event than the one in the spring, drawing Jeff Bezos, Bill Gates, and Rupert Murdoch.

There had been lots of good news for Bitcoin in the weeks since he had been in Argentina. The United States Marshals Service had auctioned off the 29,655 Bitcoins it had seized from Ross Ulbricht, and the winner was a major venture capitalist, Tim Draper, who was working with the startup that employed Nick Szabo. Once U.S. government officials had sold Bitcoins it would be hard for them to treat Bitcoin as an outlaw currency. The Winklevoss twins, meanwhile, had made their latest regulatory filing for their Bitcoin exchange-traded fund, which was now set to trade on the Nasdaq Stock Exchange under the ticker symbol COIN. The day before the Allen & Co. conference began, Wences officially announced the $20 million he had raised from Reid Hoffman, Max Levchin, and several other investors, making him the best-funded Bitcoin company in the world, according to publicly released data.

At the Allen & Co. conference, Wences was given one of the speaking slots before Jeff Bezos and Warren Buffett took the stage. Wences gave what was becoming a standard talk, beginning with the history of money, and going on to discuss the potential for Bitcoin to provide financial services to poor people who had long been shut out. He touched on Xapo only briefly, at the end. After Wences came down and took a seat with Belle, Bezos said from the stage that it was the kind of talk that kept him coming to these events.

In the hallway walking to lunch, after the Bezos-Buffett conversation, Wences spotted Bill Gates, who had been notably reticent about Bitcoin. Wences knew that Gates's multibillion-dollar foundation had been making a big push to get people in the developing world connected financially, and Wences approached him to explain why Bitcoin might help his cause. As soon as Wences broached the topic, Gates's face clouded over, and there was a note of anger in his voice as he told Wences that the foundation would never use an anonymous money to further its cause.

Wences was somewhat taken aback, but this was not the first time he had been challenged by a powerful person. He quickly said that Bitcoin could indeed be used anonymously—but so could cash. And Bitcoin services could easily be set up so that users were not anonymous. He then spoke directly to the work that Gates was doing, and noted that the foundation had been pushing people in poor countries into expensive digital services that came with lots of fees each time they were used. The famous M-Pesa system allowed Kenyans to hold and spend money on their cell phones, but charged a fee each time.

"You are spending billions to make poor people poorer," Wences said.

Gates didn't just roll over. He vigorously defended the work his foundation had already done, but Gates was less hostile than he had been a few moments earlier, and seemed to evince a certain respect for Wences's chutzpah.

Wences saw the crowd that was watching the conversation, and knew he had to be careful about antagonizing Bill Gates, especially in front of others. But Wences had another point he wanted to make. He knew that back in the early days of the Internet, Gates had initially bet against the open Internet and built a closed network for Microsoft that was similar to Compuserve and Prodigy—it linked computers to a central server, with news and other information, but not to the broader Internet, as the TCP/IP protocol allowed.

"To me it feels like you are trying to get the whole world connected with something like Compuserve when everyone already has access to TCP/IP," he said, and then paused anxiously to see what kind of response he would get. What he heard back from Gates was more than he could have reasonably hoped for.

"You know what? I told the foundation not to touch Bitcoin and that may have been a mistake," Gates said, amicably. "We are going to call you."

After Wences got back to California, he received an e-mail from the Gates Foundation, looking to set up a time to talk. Not long after that, Gates made his first public comments praising at least some of the concepts behind Bitcoin, if not the anonymity.

And so Bitcoin and its believers attracted one more person who was willing to give this new technology a look, and remain open to the possibility that the whole thing wasn't, at least, entirely crazy.

EPILOGUE

●

The summer of 2014 was the end of the first arc of Bitcoin's development. The company that been at the center of Bitcoin's initial development into a global phenomenon—Mt. Gox—had burned out, following the fate of the Silk Road and BitInstant and so many other early companies and developers that had helped take Bitcoin from nothing to something.

The dramatic crash of Mt Gox did enormous damage to the public interest and confidence in Bitcoin—more than any other previous Bitcoin related fiasco—setting the price of Bitcoin on a downward slide that continued largely unabated through the summer and fall of 2014.

What was essentially the worst thing that could happen to Bitcoin had happened. And yet Bitcoin itself had not died as had seemed all too possible when Mark Karpeles had finally admitted just how much money was gone.

The big financial institutions that had begun exploring the technology now went largely silent, and remained that way for much of 2014. Mt. Gox had made many of the these powerful companies nervous to connect themselves in any way to Bitcoin. But behind the scenes the work did not stop. Not only did Goldman Sachs and JPMorgan not disband their Bitcoin working groups, new staff members joined and other banks created their own working groups. Leaked information suggested that other corporate giants, like IBM, were exploring how they might be able to harness the new methods of record keeping provided by Bitcoin's decentralized ledger, the blockchain. They saw value in an irrevocable and attack-proof way to keep track of massive amounts of data. Wences Casares, meanwhile, stayed in touch with the Gates Foundation, which was quietly putting together a new project encouraging the use of digital money systems in the developing world, to lower the barriers that kept poor people out of the financial system. The system, it soon emerged, looked a lot like Bitcoin.

It was, though, increasingly essential to all the high powered players getting involved that the project no longer be steered by the aficionados and idealists who had been so key to getting Bitcoin through the first six years. The year 2014 marked a sharper division, than ever before, between Bitcoin's hobbyist beginnings and an increasingly professionalized future.

The Winklevoss twins, still recovering from the sting of their affiliation with Charlie Shrem, began quietly building a new Bitcoin company at a row of desks in the back of the Manhattan offices of Winklevoss Capital. Their goal was to create a new kind of Bitcoin exchange catering not to Roger Ver and Wences Casares but instead to Goldman and Citibank. As the twins began hiring for their new venture, they decided they would avoid people who had previously worked for Bitcoin companies. Instead, they began recruiting from big hedge funds and technology companies like AirBnB.

The twins were also still working to get regulatory approval for their exchange traded fund, or ETF, which would be backed by Bitcoins but trade like an ordinary stock. In the middle of 2014, after a year of work, the twins announced that the ETF would be listed on the Nasdaq Stock Exchange and carry the ticker symbol COIN if and when it was approved. The meetings at which this was decided were not run by mid-level Nasdaq executives. Instead, Nasdaq's chief executive, Robert Greifeld showed up to pitch the twins on the advantages of listing the ETF on his exchange. After the meeting, the twins began hearing that Nasdaq was not just working with them—the exchange was also dedicating its own employees to look at how the company might employ the Bitcoin technology within its core business, whatever happened with the ETF. Indeed, in Nasdaq's Manhattan headquarters, a handful of developers began working on the prototype for a system that would use something like the blockchain to execute and record the trading of stock in startups that had not gone public. The idea was that shares of a startup like Uber would be contractually tied to a fraction of a Bitcoin or something similar. When someone sold the share, the fraction of a Bitcoin would be immediately transferred to the wallet of the new owner. Every trade would be recorded on the blockchain and impossible to revoke. This seemed to come with obvious benefits over the existing system in which many startups kept track of who owned their shares in Excel spreadsheets that were sloppily kept and often inaccurate. It was clear to everyone within Nasdaq that if the experiment worked, everything on the Nasdaq stock exchange could eventually migrate on to the new platform.

FEW THINGS EPITOMIZED the fate of Bitcoin's early followers more than the trial of Ross Ulbricht, which got underway in early 2015. Ross had won the support of many early Bitcoin idealists, such as Roger Ver, who had chipped in to help pay his legal fees. On the first day in court, Ross showed up smiling and neatly groomed, with a haircut that appeared to be aimed at convincing the jury that he was an innocent suburban Boy Scout, not a criminal mastermind. Ross's lawyer, a man who had previously defended prominent Islamic terrorists, surprised everyone during his opening argument by admitting that Ross had indeed created the Silk Road back in 2011, but he said Ross had soon stepped back from the day-to-day operation of the site and thus should not be held responsible for it.

As the trial got underway there were more surprises. On the stand, federal agents acknowledged that during the summer of 2013, just months before Ross was arrested, a number of agents on the hunt had still been convinced that Mark Karpeles in Tokyo was the Dread Pirate Roberts. Ross's lawyer was eager to use this to show the flimsiness of the case against Ross. But the judge shot down that tactic and in the days that followed the main surprise was just how feeble Ross's defense really was. The laptop that the federal agents had captured when Ross was arrested contained what amounted to an archive of his guilt, documenting what Ross was thinking and doing at each stage in the Silk Road's development—and proving beyond much of a reasonable doubt that Ross had kept tight control of the operation from the beginning to the end. When the jury went to deliberate it didn't take them long to decide he was guilty on all counts.

Things got a bit more interesting a few weeks after the guilty verdict when prosecutors unsealed charges against two of the federal agents in Baltimore who had been involved in tracking down Ross—one with the Secret Service, the other with the Drug Enforcement Agency. The DEA agent, who had the impossibly colorful name Carl Mark Force IV, had been the one controlling the undercover Silk Road account nob. The new complaint explained that it was Force who had been behind many of the attempts to blackmail Ross—and it was Force who had collected those payments from Ross and cashed them out for his own financial gain, eventually extracting nearly $800,000.

It initially seemed as though the revelations might help Ross fight his conviction. After all, Ross himself had been the victim of rogue federal agents. But during the trial the government prosecutors had steered clear of introducing anything with any relation to the tainted agents, knowing that it wouldn't ultimately stand up. That made it much harder for Ross's lawyer to argue that the case presented against him in court had been compromised by the corrupt agents. Ultimately, as the judge considered what sentence to give Ross, Ross seemed to acknowledge the full extent of what he had done—if only to argue for a shorter prison term, instead of the life sentence that prosecutors were recommending. When confronted at the sentencing hearing by parents whose children had overdosed on Silk Road-purchased drugs, Ross broke down.

"I never wanted that to happen," Ross told them and the judge. "I wish I could go back and convince myself to take a different path."

The judge was not moved and gave Ross the full life sentence, explaining that it was needed to deter others from setting up similar businesses. It went unmentioned that at the time of the sentencing there were more drugs being sold online for Bitcoin than there had been on the day of Ross's arrest.

In the midst of all of this, Charlie Shrem headed to a minimum security federal prison camp in rural Pennsylvania to serve a two year sentence. Charlie had ultimately decided to plead guilty, but only to the lesser count of aiding and abetting the operation of an unlicensed money transmitting business. As he got ready to depart, he seemed to have regained much of his old spirit and verve.

"I've met with dozens of VC's and offered free advice to hundreds of companies, got to spend time with friends, throw parties and enjoy time with the love of my life," he wrote in a blog post on one of his last days of freedom. "Making the best of house arrest requires some creativity, like hiring a Jazz Duo to serenade us on Valentine's Day (bonus points because I even got to pay in Bitcoin)."

Just about the only person who seemed to be stubbornly maintaining his innocence was Mark Karpeles, who remained holed up in his apartment in Tokyo, emerging occasionally to give an interview about what a personal toll the downfall of Mt. Gox had taken on him. Meanwhile, the authorities in both Japan and the United States continued to build criminal cases against him.

The organization that Mark, Roger and Charlie had created and that had helped Bitcoin enter the mainstream—the Bitcoin Foundation—crumbled as they disappeared from the scene. While Bobby Lee tried to stabilize the organization from China, it came under attack from Bitcoin purists who hated the fact that a decentralized technology like Bitcoin had a centralized locus of power like the foundation. After one of these critics won a seat on the board, he released damning documents about the foundation's financial situation and the organization quickly lost whatever influence it had left.

Rather than empowering the ideologically-motivated wing of the Bitcoin world, though, the dissolution of the foundation paved the way for a greater professionalization of the industry. The venture capital firm Andreessen Horowitz and a few other big investors helped set up the Coin Center, in Washington, D.C. to educate and lobby regulators and legislators, and the center quickly established itself as the voice of the new elite within Bitcoin.

Gavin Andresen, who had been paid by the foundation, shifted his affiliation to the MIT Media Lab, which started its own digital currency initiative. As part of the deal, Gavin continued working from his home office in central Massachusetts. Three of the other core developers took jobs with a startup called Blockstream that was working on commercial applications for the Bitcoin blockchain.

THE METAMORPHOSIS THAT Bitcoin had been going through quietly during 2014 became known to the world in 2015.

After a year of work, the Winklevoss brothers revealed their nearly-finished Bitcoin exchange, called Gemini, or twins in Latin. Cameron and Tyler said they hoped to be the first company to get a BitLicense from New York regulator Ben Lawsky so that they could open to customers across the country. Bitcoin purists like Erik Voorhees said that the BitLicense was a betrayal of all the Bitcoin stood for and a few companies said they would rather close to customers in New York than apply for the license. But the twins were all too happy to play nice with the authorities—they believed this was what Bitcoin needed to expand.

It soon became clear, though, that the antipathy of the political purists was not the biggest problem facing Cameron and Tyler. Not long after the twins unveiled their plans, they were beat to the punch by a company that had been operating in the background. The company, itBit, had originally opened a Bitcoin exchange in Asia, but the former hedge-fund traders who owned it announced in May of 2015 that they had put together a high-powered board to expand to the United States, including the former chairwoman of the Federal Deposit Insurance Corporation, Sheila Bair, and the former presidential candidate and basketball player Bill Bradley. Even more impressive, itBit had gotten approval from New York regulators for something that was more useful than a BitLicense—a trust company charter—which allowed it to open as a full-service Bitcoin exchange with a fully-functioning American bank account. This was the elusive dream that Mt. Gox had been chasing as far back as 2012.

Even itBit, though, quickly found itself on the back foot as more and more of the work that had been going on behind the scenes during the previous year came to fruition. The big banks came out one at a time to announce the work they had been doing in the area. Some of them talked about their interest in Bitcoin itself. Goldman Sachs, for instance, led an investment in a Boston-based startup, Circle, that aimed to use Bitcoin as the rails for a new payment system that would allow young people to send money to the mobile phones of their friends, similar to the service that startups like Venmo offered.

But most of the bankers explained that what they were not interested in Bitcoin and they certainly weren't looking to trade virtual currencies on an exchange like itBit or Gemini. What captivated them was the blockchain technology that Bitcoin had introduced. Executives from Citibank, Santander and BBVA, among others, said that they saw no limit to the potential of decentralized ledgers that enabled cheaper, faster financial transactions of all sorts. UBS and Barclays set up dedicated blockchain laboratories in London where they experimented with ways to link up with other banks on blockchain-like systems to conduct large transactions. The projects were somewhat like the software that Nasdaq had been building to record and execute stock trades, but now there was talk of using blockchain of different sorts to handle every sort of financial transaction in the world.

Some of the systems being discussed were tied into the Bitcoin blockchain, but most banks wanted to steer clear of anything related to Bitcoin. Wall Street still associated the virtual currency with the Silk Road and illegal transactions. Financial executives were also put off by the Bitcoin mining free-for-all, which was open to anyone with a computer and an internet connection. Bitcoin mining operations in China had become ever more powerful thanks to ready availability of cheap computer hardware and electricity. Over half of the new Bitcoins being minted each day were going to Bitcoin mining farms in China. This terrified the banks, who knew that the miners served as the accountants on the Bitcoin network. Goldman Sachs had no interest in having any part of its business reliant on anonymous, far-off computer farms. There were, not surprisingly, plenty of entrepreneurs willing to provide Goldman and other banks with blockchain-like technology that could operate independent of Bitcoin. The most high profile of these new companies had at its helm Blythe Masters, a storied JPMorgan banker who was given credit for inventing the credit default swap, a financial product that was at the center of the financial crisis. At a gathering with Wall Street executives in the summer of 2015, Masters said: "You should be taking this technology as seriously as you should have been taking the development of the Internet in the early 1990s."

The notion that the concept of the blockchain could somehow be separated from Bitcoin the virtual currency seemed silly to many of the long-time Bitcoin backers in Silicon Valley. It was Bitcoin, after all, that had provided the incentives for new people to keep the blockchain and secure it against attacks. A blockchain that was maintained by just 20 banks was much easier to overpower. Wences Casares asked how a private blockchain, accessible only to a limited group, was any different than an old-fashioned database. Wences liked to say that the desire to use the blockchain without Bitcoin was like somebody wanting to have a website without using the internet. Bitcoin was the thing that made the blockchain work, Wences argued.

The debate over this supposition—that a powerful blockchain required a virtual currency like Bitcoin—dominated the increasingly frequent industry events where Wall Street executives came together to discuss their work in the area. These events also drew attendees from central banks, including the Federal Reserve and the Bank of England, who described their own interest in blockchains, including the possibility that their own currencies would eventually end up being created and recorded on some sort of decentralized ledger. They loved the idea that all of the money would be recorded in a single database that everyone could access through the wonders of cryptography.

Whatever the disagreements about how it would happen, both the Bitcoin and blockchain supporters seemed to agree on the fact that this technology was going to fundamentally change industries around the world, and not just finance. In an essay entitled "Blockchain: It Really is a Big Deal," the head of research at IBM, Arvind Krishna, wrote that blockchain technology was set to change any industry that required trust between two parties and that relied in the current world on some middleman to broker transactions, including law and international trade.

"It has the potential to vastly reduce the cost and complexity of getting things done," he wrote. "Essentially, it could help bring to business processes the openness and hyper efficiency we have come to expect in the Internet Era."

In a survey conducted by the World Economic Forum, business leaders predicted that the first governments would begin collecting taxes on some version of the blockchain in 2023 and that 10 percent of the world's gross domestic product would be stored on some form of a blockchain by 2027.

At events—like one at the New York Times conference center, sponsored by Citi—Bitcoin old-timers were awestruck by all the folks involved in the conversation who wouldn't have been willing to go anywhere near a Bitcoin event a year earlier.

It was less often remarked upon that the institutions that seemed to be driving the conversation—Wall Street and central bankers—were the very centers of power that Bitcoin had been designed to disempower and circumvent back in 2008.

EVEN WITH THE blockchain dominating the conversation, good old fashioned Bitcoin continued soldiering along. There were still plenty of people speculating on the price of Bitcoin—especially in China—and using the virtual currency to do things online that weren't possible with credit cards, most of them illegal. That—and the excitement about the blockchain—helped push the price of Bitcoin above $200 and gave an incentive to miners to keep this supercharged network of computing power going. But there still didn't seem to be much pickup among ordinary consumers, particularly in America, where there were, as yet, few advantages to using Bitcoin instead of a credit card or PayPal. The big companies that had decided to accept Bitcoin had seen the number of transactions trail off. Wences Casares had largely given up on the American market for his company, Xapo, and was going all in on a strategy of catering to people in emerging economies like India, where people otherwise lacked access to the financial system. Xapo was building systems to allow people in India to buy small amounts of Bitcoin to access the world of online commerce that was otherwise closed off to them. The question was whether the money that Wences had raised would be enough to keep his company going until the numbers really picked up—if they really picked up.

Even down at the most basic level of the Bitcoin protocol—now in version 0.11—there were fundamental questions about whether Bitcoin could live up to the big ambitions that people had been dreaming up since Hal Finney downloaded version 0.1.

Gavin Andresen was still dedicating his days to the project, but he worried that the Bitcoin protocol was going to run into a wall. The original software had limited the number of transactions that could be processed every 10 minutes to about 7 transactions a second—or 1 megabyte of data in each block that the miners confirmed. That had been fine so far, but if banks were going to be using the blockchain to process transactions the system would need to handle much bigger blocks of data. The Visa network was processing over a thousand transactions a second.

Gavin had been pushing the four other core developers of the Bitcoin protocol to approve changes. Gavin teamed up with Mike Hearn, the Switzerland-based developer who had been emailing with Satoshi back in early 2009. They also had the support of big Bitcoin companies like Coinbase and Circle, which wanted a network with the capability to grow.

But three of the other core developers were not interested in the changes Gavin was proposing, and there was an unwritten agreement to not move ahead on new versions until there was a consensus. The terms of the debate were technical, but it came down to basic disagreements about what the Bitcoin network should look like. Gavin worried that if only 7 transactions a second we're going through, miners would begin to charge larger and larger fees to include a transaction in a block and make it official. The only people who would be willing to pay these fees, as Gavin saw it, would be large companies with large transactions. This didn't seem to live up to the open access that Satoshi had written about.

On the other side, the three core developers, all of whom were now associated with the startup Blockstream, worried that if more transactions were coursing through the system it would be hard for ordinary computers to process the data and participate in the decentralized network. Soon enough, they said, only a few industrial scale miners would be handling all the data. That sort of centralized control scared the Blockstream developers and seemed too close to the concentration of power in the existing financial systems, with all the problems that had brought.

The debate went back to the core questions about the dangers of centralization and what sort of centralization Bitcoin had been set up to fight. Gavin and Mike Hearn wanted to keep it easy for anyone to conduct transactions on the network while the Blockstream developers wanted to make it easier for anyone to process transactions, in order to keep the network itself decentralized. Both sides referred back to Satoshi.

The debate got surprisingly personal, especially after Gavin and Hearn released their own proposed software, Bitcoin XT, which included the changes they wanted to see. They said that they wanted to allow the people using Bitcoin to vote on the change. If more than 75% of the miners started using the software it would become the official protocol. Opponents of Gavin and Mike accused them of being 'dictatorial' and betraying the culture of consensus that had prevailed so far.

The Blockstream developers, meanwhile, were accused of working for the interests of their company, Blockstream, rather than the good of Bitcoin as a whole. Their company, after all, proposed breaking many smaller transactions off onto so called sidechains that would only occasionally link back up with the main Bitcoin blockchain, making broad access to the network less important. But the Blockstream developers ultimately won over enough miners to their way of thinking that the Bitcoin XT software died on the vine.

The outcome confirmed some of the worst fears about Bitcoin's open source foundation. With no one in charge it was hard to make basic changes. The whole thing left Gavin discouraged and caused him to pull away from his work with the other core developers. In January of 2016, Mike Hearn announced that he had sold his last Bitcoins and was done with the project all together; he had taken a job with one of the start-ups trying to provide blockchain-solutions to Wall Street. In a blog post entitled, "The resolution of the Bitcoin experiment," Hearn wrote that "Bitcoin has gone from being a transparent and open community to one that is dominated by rampant censorship and attacks on bitcoiners by other bitcoiners."

There was, though, too much at stake, for most other people in this new industry to give up. The various mining companies and Silicon Valley startups worked furiously behind the scenes in an effort to find a way forward. The process of changing the Bitcoin protocol was a clunky process that moved slowly. But it moved. And as it did it allowed input from all the different camps who were interested in Bitcoin rather than any single source of authority. It was a conversation, and a conversation that was continuing on much longer than almost anyone had thought possible.

TECHNICAL APPENDIX

●

ADDRESSES AND SECRET KEYS

Anyone joining the Bitcoin network can generate his or her own Bitcoin address (generally a string of thirty-four letters and numbers), and a corresponding private key (generally a string of sixty-four characters).

As an example, one actual Bitcoin address is:

16R5PtokaUnXXXjQe4Hg5jZrfW69fNpAtF

The private key for this particular address is:

5JJ5rLKjyMmSxhauoa334cdZNCoVEw6oLfMpfL
8H1w9pyDoPMf3

Only the person with this private key can sign off on transactions from that address (the address is empty so don't bother trying).

Each Bitcoin address has one and only one private key. The relationship between the private key and the address is determined by a series of complex math equations, which makes it essentially impossible to work backward from the public Bitcoin address to find the private key.

A Bitcoin user can generate endless numbers of Bitcoin addresses and private keys. There is no cost for doing so. The length of the addresses and the sheer number of potential addresses ensure that it is all but impossible for the same address to be generated twice.

INITIATING A TRANSACTION

With a private key, a user, let's call her Alice again, can send money from her address without ever sharing the private key with anyone else. Rather than sending out her private key, Alice puts her private key into software on her own computer, along with details of her transaction. Without sending this information to the network, the Bitcoin software on Alice's computer runs the information through a series of complicated math equations that spits out a special code, often referred to as a digital signature. This part of the process can happen even if Alice's computer is offline. It is this digital signature—a unique product of her private key and the transaction taking place—that Alice sends out to the network along with her transaction, much like a signature on a check.

VERIFYING TRANSACTIONS

The computers that get Alice's digital signature are unable to work backward to get Alice's private key, thanks to the mathematical innovations involved. But the computers can put Alice's digital signature and her public Bitcoin address into another series of complicated math equations and verify that the digital signature was, indeed, created by the private key corresponding to the public address. Again, these are very sophisticated mathematical manipulations that happen on both sides of this, on one side to generate the signature and on the other to verify it.

It is necessary for the computers on the network to verify every transaction because there is no central authority to do this work. Once the computers do verify that Alice has the right private key, they then check that Alice's Bitcoin address has the coins she is trying to send. The computers on the network do this by scanning the record of all previous Bitcoin transactions coming to and from the address Alice is using.

CREATING BLOCKS AND RECORDING TRANSACTIONS (THE BITCOIN MINING PROCESS)

Satoshi saw that it would be problematic if each computer on the network recorded every transaction as it arrived. A transaction might reach one computer before it reached another computer on the network, leading to disagreements about the balance in each address. Bitcoin needed to have one definitive record of when each transaction occurred, and Satoshi came up with a clever way to achieve this through the use of a kind of ongoing contest that any member of the network could compete in.

To win the contest, all the computers on the network would compile recent transactions, as they were sent around the network, into long lists, which were referred to generically as blocks. After compiling the transactions into a block, a computer would then run the block through yet another specialized math equation, known as a hash function, which can take any data—the Gettysburg Address or your name—and turn these data into a unique sixty-four-character digest. The computers taking part in the Bitcoin contest are looking for a block that can be put into a hash function known as SHA 256 and generate a sixty-four-character digest with a specific number of zeroes at the beginning. If, for

instance the computers are looking for a digest with five zeroes at
the beginning, either of these digests would be a winner:

    000006d77563afa1914846b010bd164f395bd
    34c2102e5e99e0cb9cf173c1d87

Or

    000007ac6b77f49380ea90f3544a51ef0bfbfc
    8304816d1aab73daf77c2099319

Because SHA 256, like other hash functions, is essentially impossible to reverse-engineer, it is impossible to tell what sort of
block will lead to a digest with five zeroes at the beginning.

Given that SHA 256 and other hash functions always generate
the same digest from any particular input, if every computer put the
same transactions into their block, every computer would get the
same digest out the other end. In order to differentiate their blocks,
in the hope of finding a winning block, each computer would be
tasked with adding a random number onto the end of the block.
Because of the sensitive nature of hash functions, changing the
random number at the end of the block from 20 to 22 could potentially change the digest from a digest with one zero to a digest
with ten zeroes at the beginning. If one random number didn't lead
to a digest with the desired number of zeroes, the computer would
try the block with another random number attached to see if that
worked. All the computers hoping to win would keep trying out
new random numbers—and adding incoming transactions—until
one computer found a block that led to a digest with the correct
number of zeroes. Because finding an answer involved trying out
random numbers, this contest was more a game of luck than a
game of skill—but the computer that could run guesses through

the hash function fastest would increase its chances of winning, just as a person with twenty lottery tickets has a better chance of winning than a person with only one.

The number of zeroes required to win the contest was somewhat inconsequential but made it easy to adjust the difficulty of the contest and ensure that new blocks arrived approximately every ten minutes. If computers were winning more often than every ten minutes, the Bitcoin software could adjust and demand that computers find a digest with more zeroes at the beginning. If computers were not winning frequently enough, the software could adjust and allow winners to have less zeroes. As the contest became harder, it required more high-powered computer hardware to win it.

WINNING BLOCKS

When a computer did find a winning block, it would send the winning block around the network, so that the other computers could verify that the block did indeed generate a digest with the desired number of zeroes at the beginning. The computers would then add the winning block to the blockchain held on all the computers, thus recording the list of transactions included in the block. That block became the official record of all transactions that occurred since the previous winning block. If the winning block left out a few transactions that were included in the blocks created by other computers, those transactions would not be recorded on the blockchain and would be left out for the next round of blocks. In addition to the transactions and the random number, the blocks also included a reference to the previous block and data on the state of the Bitcoin network, so that all this information would also be recorded on the blockchain.

The creative method for arriving at a single, communally agreed upon record of transactions provided a long-sought solution to a conundrum known as the Byzantine Generals Problem. Before Bitcoin, computer scientists struggled with how to build a reliable network of unrelated people, if some of the people could not be trusted. The method of building a blockchain, with each block coming from just one member of the network, and disagreements being solved by majority rule, solved this problem.

GENERATING NEW COINS

When a computer generated a winning block, it also won a bundle of new coins—50 Bitcoins when the system first began. These coins were created in a clever way. In essence, when computers were generating the list of transactions in a block, they included, in their list of transactions, a transaction granting one of their own Bitcoin addresses 50 Bitcoins out of thin air. When a block won the lottery, and was added to the blockchain, this seemingly fictional transaction was turned into a reality, and the address in question had 50 more Bitcoins attached to it. By making it onto the blockchain the transaction was made real. The transaction that created new Bitcoins would be referred to as the coinbase of each block. If a computer tried to grant itself more than 50 new Bitcoins, the whole block would be rejected by the other computers, even if it generated a digest with the correct number of zeroes.

ACKNOWLEDGMENTS

●

L ike Bitcoin, this book was an act of group invention made possible by many wonderful people. Andrew Ross Sorkin brought me into the job that allowed me to start writing about this fascinating topic. Later on he saw that there was a bigger story to be written about Bitcoin and pushed me to write it. I can't thank him enough. My agent, Andrew Wylie, gave me the confidence I needed to take this idea out into the world and find it the right home. At HarperCollins, Tim Duggan immediately understood what I was hoping to do with this book, and Jonathan Jao made sure I did it. Both of them were the kind of editor every young writer dreams of finding. Emily Cunningham was my guide and good fairy through the entire process. I am also grateful for the help I was given by Joanna Pinsker, Stephanie Cooper, and the rest of the staff at HarperCollins.

This book is, at its core, the story of several people who opened up their lives to me. I have to thank, most of all, Wences Casares, Barry Silbert, Bobby Lee, Charlie Shrem, Roger Ver, Martti Malmi, Gavin Andresen, and Tyler and Cameron Winklevoss.

But the story wouldn't have come together without the time and cooperation of Fran, Hal, and Jason Finney; Dan Morehead; Patrick Murck; Erik Voorhees; Jesse Powell; Mark Karpeles; Mike Hearn; Naval Ravikant; Jed McCaleb; MiSoon Burzlaff; Nick Szabo; Reid Hoffman; Eric O'Brien; Federico Murrone; Charlie Lee; Amir Taaki; Jamileh Taaki; Alex Rampell; Emmauel Abiodun; Nicolas Cary; David Marcus; Jorge Restrelli; Bill Tanona; Pete Briger; Jamie Dimon; Max Neukirchen; Andy Dresner; Paul Walker; Marty Chavez; Alexander Kuzmin; Nicole Navas; Lyn Ulbricht; Josh Dratel; John Collins; Jennifer Shasky Calvery; Sebastian Serrano; Chris Larsen; Chris Dixon; Balaji Srinivasan; Marc Andreessen; Kim Milosevic; Brian Armstrong; Fred Ehrsam; John O'Brien; Belle Casares; Patrick Strateman; Yifu Guo; Marcie Braden; Alex Waters; Brian Klein; Nejc Kodric; Paul Chou; Jeff Garzik; Adam Back; Laszlo Hanecz; Leon Li; Gil Lauria; Monica Long; Michael Keferl; Daniel Kelman; Jack Smith; Tim Swanson; Rui Ma; Jack Wang; Ling Kang; Huang Xiaoyu; Kathleen Lee; Ayaka Ver; Alex Likhtenstein; Jeremy Allaire; Matt Cohler; Larry Lenihan; Fred Wilson; Michael Goldstein; Phil Zimmerman; Yin Shih; Perry Metzger; Tony Gallipi; Bruce Wagner; and Justin Myers. I also was lucky to be writing about a topic that had already been covered by smart journalists, academics, and filmmakers like Nicholas Mross, Joshua Davis, Kevin Roose, Eileen Ormsby, Izabella Kaminska, Felix Salmon, Andy Greenberg, Sergio Demian Lerner, Sarah Meikeljohn, Nicolas Christin, Susan Athey, Adrianne Jeffries, and Andrea Chang.

This book immensely benefited from my first readers, some of whom are also my best friends: Teddy Wayne, Peter Eavis, Lev Moscow, Mark Suppes, David Segal, Benny Gorlick, Alex Morcos, and Ben Davenport. My friends Danielle and Alex Mindlin, and Gal Beckerman and Deborah Kolben gave me lots of good advice and listened to my griping. Mirta Kuperminc and her family graciously put me up while I did my work in Argentina.

I'm lucky to work for the *New York Times* and DealBook, where the exceptional staff make it exciting to go to the office each day. In my time at the paper, Arthur Sulzberger Jr., Jill Abramson, and Dean Baquet have kept the paper dedicated to the ideals that made it a place I wanted to work for from the time I became a journalist. Several wonderful editors helped me develop my ideas and put up with my absence while I developed them into a book. They include Jeffrey Cane, Dean Murphy, Vera Titunik, David Gillen, and Peter Lattman, who brought me into my very first Bitcoin story. My colleagues Charles Duhigg, Jim Stewart, Ron Lieber, Barry Meier, and David Gelles shared wisdom that made it a bit easier to navigate the book-writing process for the first time. I am also forever indebted to the editors and journalists who gave me a shot at various points in my career and helped me grow. The list begins with J.J. Goldberg and extends to Ami Eden, Alana Newhouse, John Palattella, Geraldine Baum, Davan Maharaj, Tom Petruno, and Larry Ingrassia, among others.

This book was, in the end, possible only because of my family: Lewis, Sally, and Miriam Popper; Juliana, Robbie, Florence, and Beatrice Dapice; and my broader family, the Strauss clan, with special thanks to Jona, Martin, and Alanna, who helped care for my family when I could not. My son, August, put up with too little time with his father and gave me an incentive to finish. My beloved wife, Elissa, did everything that no one else could do for me, and more, allowing me to accomplish things that would be impossible without her.

SOURCES

The bulk of this book is based on over three hundred interviews I conducted with the people involved, in places as far flung as Buenos Aires; Beijing; Shanghai; Tokyo; Austin; San Francisco; Palo Alto; Reykjavik; Toronto; Washington, DC; Amsterdam; and New York. I was often able to confirm the recollections with private e-mails and other contemporaneous documents that were shared with me. In the end only a handful of the people mentioned in this book declined to talk to me.

Unless I have specified otherwise in the notes below, readers can assume that every moment described in this book came to me directly from at least one or, when possible, more than one person present at the event described. Most of the direct quotes come from contemporaneous documents or recordings but some of the quotes are the best recollection of the participants, generally backed up by at least one other person in attendance. I was lucky enough to be present for some of the events, such as the March 2014 gathering at Dan Morehead's house on Lake Tahoe.

Most of the material that did not come from interviews and personal e-mails sat in the digital treasure trove of public messages and chats that the Bitcoin community has created over time, and that various participants had the wisdom to maintain for posterity. They will be referenced in the notes by following abbreviations:

CYPH: Cypherpunk mailing list, http://cypherpunks.venona.com/.
CRYP: The Cryptography and Cryptography Policy Mailing List, http://
www.mail-archive.com/cryptography@metzdowd.com/.
DEV-LIST: Core Bitcoin development discussion, http://sourceforge
.net/p/bitcoin/mailman/bitcoin-development/.
BTCF: Bitcoin Forum, https://bitcointalk.org.
IRC: #bitcoin-dev Internet Relay Chat channel, http://bitcoinstats.com/
irc/bitcoin-dev/logs/2014/01.

On Silk Road, there are two remarkable online efforts to gather and cat-
alog all available information, including legal documents and postings from
the now defunct marketplace. One is available at http://antilop.cc/sr/. The
other is at http://www.gwern.net/Silk%20Road. Many of the details in the
book came from the Silk Road's forums and Ross Ulbricht's trial, which will
be referred to in the notes by the following abbreviations:

SRF: Silk Road forum archives, http://antilop.cc/sr/download/stexo_sr_
forum.zip.
RUTT: Ross Ulbricht trial transcripts, *United States of America v. Ross
William Ulbricht.* United States District Court Southern District of New
York. 14 CR 68 (KBF).
RUTE: Ross Ulbricht trial exhibits, *United States of America v. Ross
William Ulbricht.* United States District Court Southern District of New
York. 14 CR 68 (KBF).

The notes below will not contain citations for material from the sources
above when it is obvious in the text where the material came from.

All Bitcoin prices are taken from CoinDesk's Bitcoin Price Index, which is
available at http://www.coindesk.com/price/, unless I have stated otherwise.
The numbers on Bitcoin trading volumes come from www.bitcoinmarkets.
com and www.bitcoinity.com/data.

For those looking to learn more about the topics covered in this book
there are several wonderful books. On the history of the Cypherpunks,
there is Andy Greenberg's *This Machine Kills Secrets: How WikiLeakers,
Cypherpunks, and Hacktivists Aim to Free the World's Information.* For
the history of cryptography I learned a great deal from Simon Singh's *The
Code Book.* For those eager to learn more about the evolution of money,
Felix Martin's *Money: An Authorized Biography* and Jack Weatherford's
The History of Money are wonderful reads, and Nigel Dodd's *The Social*

Life is thought-provoking. Those looking to go into greater depth can try *A History of Money* by Glyn Davies. I also benefited from Eileen Ormsby's book *Silk Road*, the first of what I'm sure will be many fascinating volumes about the online bazaar.

INTRODUCTION

xiv only 15 percent of the basic Bitcoin computer code: Based on calculations done for the author by Gavin Andresen.

CHAPTER 1

4 this particular e-mail came from: Satoshi Nakamoto to CRYP, October 31, 2008.

4 the nine-page description: A later version of the paper would be nine pages, but the initial version Hal reviewed was actually eight pages.

5 tied to an Internet provider in California: Hal's debug log showed that the IP addresse of the other user was reached through a Tor service that would have obscured the real IP address. But Tor generally routes users to nodes in the same geographic area, suggesting that the other user on Bitcoin's first day was probably in California.

5 He said he'd been testing it heavily: I have elected to use the pronoun "he" to refer to Satoshi, but Satoshi could also be she or they.

6 now recorded next to one of his Bitcoin addresses: The address in question was 1AiBYt8XbsdyPAELFpcSwRpu45eb2bArMf.

12 Chaum's effort would rub Hal and others the wrong way: Hal Finney to CYPH, August 22, 1993.

12 DigiCash went down with it: Tim Clark, "DigiCash Files Chapter 11," CNET, November 4, 1998, http://news.cnet.com/2100-1001-217527.html.

13 Hal would calculate the maximum bill: This anecdote was recounted by Hal's college roommate and later colleague, Yin Shih.

13 "The work we are doing here, broadly speaking": Hal Finney to CYPH, November 15, 1992.

CHAPTER 2

16 As sociologist Nigel Dodd put it: Nigel Dodd, *The Social Life of Money* (Princeton, NJ: Princeton University Press, 2014).

17 "We could envisage proposals in the near future": Alan Greenspan, Conference on Electric Money and Banking, United States Treasury, September 19, 1996, http://www.federalreserve.gov/boarddocs/speeches/19960919.htm.

17 a British researcher named Adam Back released his plan: Adam Back to CYPH, March 28, 1997.

18 a concept called bit gold, was invented by Nick Szabo: Nick Szabo, "Bit Gold," *Unenumerated*, December 2005, http://unenumerated.blogspot .co.uk/2005/12/bit-gold.html.

19 Another, known as b-money, came from an American named Wei Dai: Wei Dai to CYPH, 1998.

19 Hal created his own variant, with a decidedly less sexy name: Hal Finney to CYPH, August 15, 2004.

20 The nine-page PDF attached to the e-mail: the current version is available at https://bitcoin.org/bitcoin.pdf.

22 modeled after the contest that Adam Back: While this process was modeled on Back's program, it also relied on the innovations of several other cryptographers and mathematicians, including Ralph Merkle, Stuart Haber, and W. Scott Stornetta.

25 usually belonging to Satoshi: Satoshi's mining activities were traced by the Argentinian researcher Sergio Demian Lerner. Sergio Demian Lerner, "The Well Deserved Fortune of Satoshi Nakamoto, Bitcoin Creator, Visionary and Genius," Bitslog, April 17, 2013, https://bitslog.wordpress .com/2013/04/17/the-well-deserved-fortune-of-satoshi-nakamoto/.

25 the first transaction took place when Satoshi sent Hal ten coins: Satoshi's address for this transaction was 12cbQLTFMXRnSzktF kuoG3eHoMeFtpTu3S; Hal's was 1Q2TWHE3GMdB6BZKafqwxX tWAWgFt5Jvm3.

26 Satoshi was using his own computers to help power the network: Lerner.

26 When a programmer in Texas wrote to Satoshi late one night: The programmer, Dustin Trammel, posted the e-mails on his blog at http:// blog.dustintrammell.com/2013/11/26/i-am-not-satoshi/.

CHAPTER 3

29 Before reaching out to Satoshi, Martti had written about Bitcoin on anti-state.com: Martti's post, written under the screen name Trickster, is available at https://board.freedomainradio.com/topic/17233 -p2p-currency-could-make-the-government-extinct/.

30 "The root problem with conventional currency": Satoshi Nakamoto, "Bitcoin Open Source Implementation of P2P Currency," P2P Foundation forum, February 11, 2009, http://p2pfoundation.ning.com/forum/ topics/bitcoin-open-source.

33 It also meant that Satoshi's computers were still: Sergio Demian Lerner,

"The Well Deserved Fortune of Satoshi Nakamoto, Bitcoin Creator, Visionary and Genius," Bitslog, April 17, 2013, https://bitslog.wordpress.com/2013/04/17/the-well-deserved-fortune-of-satoshi-nakamoto/.

35   "Be safe from the unstability caused by fractional reserve": An archived version of the page designed by Martti is available at http://web.archive.org/web/20090511173000/http://bitcoin.sourceforge.net/.

35   A few dozen people downloaded the Bitcoin program: Data on software downloads available at http://sourceforge.net/projects/bitcoin/files/stats/timeline.

37   Starting in August, the log of changes to the software: The history of changes to the software is available at https://gitorious.org/bitcoin/bitcoind/activities.

37   When the next version of Bitcoin, 0.2: Satoshi Nakamoto to DEV-LIST, December 17, 2009.

37   the majority of coins were still: Lerner.

37   throughout 2009 no one else was sending or receiving: Data on the number of transactions per block available at https://blockchain.info/charts/n-transactions-per-block.

38   In the very first recorded transaction of Bitcoin for United States dollars: Information on the transaction is available at https://blockchain.info/tx/7dff938918f07619abd38e4510890396b1cef4fbeca154fb7aafba8843295ea2.

38   NewLibertyStandard came up with his own method: The shuttered exchange is still online at http://newlibertystandard.wikifoundry.com/page/Exchange+Rate.

39   Swap Variety Shop on his exchange website: The shuttered shop is still online at http://newlibertystandard.wikifoundry.com/page/Specialty+Shop.

CHAPTER 4

44   But on May 22, 2010, a guy in California offered to call Lazlo's local Papa John's: Information about the Bitcoin transaction is available at https://blockchain.info/tx/a1075db55d416d3ca199f55b6084e2115b9345e16c5cf302fc80e9d5fbf5d48d.

44   small item on the website of *InfoWorld*: Neil McAllister, "Open Source Innovation on the Cutting Edge," *InfoWorld*, May 24, 2010, http://www.infoworld.com/article/2627013/open-source-software/open-source-innovation-on-the-cutting-edge.html.

47   "Slashdot with its millions of tech-savvy readers": Martti Malmi to BTCF, June 22, 2010.

48 "How's this for a disruptive technology?": "Bitcoin Releases Version
    0.3," *Slashdot*, July 11, 2010, http://news-beta.slashdot.org/story/
    10/07/11/1747245/bitcoin-releases-version-03.

CHAPTER 5

49 The number of downloads would jump from around three thousand:
    Data on software downloads available at http://sourceforge.net/
    projects/bitcoin/files/stats/timeline.

49 "Over the last two days of Bitcoin being": Gavin Andresen to BTCF,
    July 14, 2010.

53 the difficulty of mining new Bitcoins jumped 300 percent: Data on
    mining difficulty available at https://blockchain.info/charts/difficulty?
    timespan=all&showDataPoints=false&daysAverageString=1&show_
    header=true&scale=0&address=.

54 In one month, the forum had gained more new members: Data on forum
    usage available at https://bitcointalk.org/index.php?action=stats.

57 "Nobody can stop the Bitcoin system": Keir Thomas, "Could the
    Wikileaks Scandal Lead to New Virtual Currency?" *PCWorld*, De-
    cember 10, 2010, http://www.pcworld.com/article/213230/could_
    wikileaks_scandal_lead_to_new_virtual_currency.html.

CHAPTER 6

65 "To tell the truth, I always felt": Mark's blog has been taken down,
    but an archived version of this post is available at http://web.archive
    .org/web/20140302234940/http://blog.magicaltux.net/2006/02/12/
    pensees-nocturnes/.

69 begun in earnest in July 2010 when he had sold a cheap house in Penn-
    sylvania: RUTE GX 250 and GX 251.

69 Ross rented a cabin about an hour from his home in Austin, Texas:
    RUTE GX 240A.

70 he knew he wanted to set up a new kind of online market: RUTE GX
    240A.

70 His curiosity about and penchant for the outdoors: Ross spoke about
    his youth in a recording done for the StoryCorps project with his friend
    Rene Pinnel in 2012.

70 At Penn State, he had the unique distinction: Erin Rowley, "Caribbean
    Students Host Cultural Event," *Daily Collegian*, March 24, 2008,
    http://www.collegian.psu.edu/archives/article_ef9c02f3-a9c2-5b8f
    -b1d3-f0ef82e3dce0.html. Katharine Lackey, "Paul to Visit PSU," *Daily*

Collegian, March 26, 2008, http://www.collegian.psu.edu/archives/article_239513a3-a577-5732-bab0-9cc27c5d4610.html.

70 "Everywhere I looked I saw the State": Dread Pirate Roberts to SRF, March 20, 2012.

70 Initially, he called the project Underground Brokers: RUTE GX 240A.

71 he soon had big black trash bags full of them: Richard Bates, RUTT, January 22, 2015.

72 "either don't want the spouse to see it on the bill": Satoshi Nakamoto to BTCF, September 23, 2010.

73 "I felt ashamed of where my life was": RUTE GX 240A.

73 he had, by his own accounting, gone through $20,000: RUTE GX 250.

73 By the end of February, twenty-eight transactions: silkroad to BTCF, March 1, 2011.

CHAPTER 7

75 "i'm so stressed! i gotta": Richard Bates, RUTT, January 22, 2015.

75 "*Free Talk Live*, who was broadcasting live at the time": FreeTalk Live, March 16, 2011, https://www.freetalklive.com/content/podcast_2011_03_16.

76 "my site had a 40 minute spot on a national": RUTE GX 1002.

77 he was sentenced to ten months in prison: Information on the case is available at http://www.justice.gov/criminal/cybercrime/press-releases/2002/verPlea.htm.

80 "Law-abiding citizens can carry on their affairs": Jerry Brito, "Online Cash Bitcoin Could Challenge Governments, Banks," Techland blog, *Time*, April 16, 2011.

80 "cuts across international boundaries, can be stored": Andy Greenberg, "Crypto Currency," *Forbes*, April 20, 2011, http://www.forbes.com/forbes/2011/0509/technology-psilocybin-bitcoins-gavin-andresen-crypto-currency.html.

82 "This was—of course—denied": Mark Karpeles to BTCF, May 1, 2011.

83 Silk Road now had over a thousand people registered: Eileen Ormsby, *Silk Road* (Sydney: Pan Macmillan Australia, 2014).

83 "Updating a live site to a whole new version is no easy task": RUTE GX 240B.

83 Gawker published an in-depth story about Silk Road: Adrian Chen, "The Underground Website Where You Can Buy Any Drug Imaginable," Gawker, June 1, 2011, http://gawker.com/the-underground-website-where-you-can-buy-any-drug-imag-30818160.

83   over a thousand new people were registering for Silk Road: Ormsby.

84   "online form of money laundering used to disguise": "Schumer Pushes to Shut Down Online Drug Marketplace," June 5, 2011, http://www.nbcnewyork.com/news/local/Schumer-Calls-on-Feds-to-Shut-Down-Online-Drug-Marketplace-123187958.html.

85   earning $17,000 from the sale of his mushrooms, and $14,000 from commissions: RUTE GX 250.

85   "I was mentally taxed, and now I felt extremely vulnerable": RUTE GX 240B.

86   15,000 new people joined the forums: Data on forum usage available at https://bitcointalk.org/index.php?action=stats.

86   He said he had long avoided determining: Martti Malmi to BTCF, June 11, 2011.

CHAPTER 8

90   The selling continued until 260,000 Bitcoins were purchased: IRC, June 19, 2011.

95   appeared briefly, via Skype, on *The Bitcoin Show*: Episode 25, June 19, 2011, https://www.youtube.com/watch?v=Ye_81RH6wiI.

95   "Ready guys?": An archived version of this chat is available at http://pastebin.com/d7vp06hL.

96   "it's likely to go the way of other": Peter Cohan, "Can Bitcoin Survive, Is It Legal?" *Forbes*, June 28, 2011, http://www.forbes.com/sites/petercohan/2011/06/28/can-bitcoin-survive-is-it-legal/.

CHAPTER 9

97   the founder of a small Polish Bitcoin exchange, Bitomat, announced: Kyt Dotson, "Third Largest Bitcoin Exchange Bitomat Lost Their Wallet, Over 17,000 Bitcoins Missing," *Silicon Angle*, August 1, 2011, http://siliconangle.com/blog/2011/08/01/third-largest-bitcoin-exchange-bitomat-lost-their-wallet-over-17000-bitcoins-missing/.

98   The founder of the site, a man who called himself Tom Williams, was unresponsive: Adrianne Jeffries, "Search for Owners of MyBitcoin Loses Steam," BetaBeat, *New York Observer*, August 19, 2011, http://observer.com/2011/08/search-for-owners-of-mybitcoin-loses-steam/.

102   "I know for sure attendees are flying in": Bruce Wagner to BTCF, July 27, 2011.

104   "You can call me an idiot and yeah": Gavin's presentation is viewable at https://www.youtube.com/watch?v=0ljx4bbJrYE.

104 "be making a HUGE HUGE HUGE announcement at the Conference": Bruce Wagner to BTCF, August 14, 2011.

104 "If that's not enough": Wagner's presentation is viewable at https://www.youtube.com/watch?v=pv0SdUNcBKc.

CHAPTER 10

110 The announcement from the Free State Project: Erik Voorhees to BTCF, October 8, 2011.

111 The people who had been attending the New York Bitcoin Meetup: Disposition to BTCF, October 4, 2011.

112 "the sanctity of the individual, the priority": Mark Lilla, "The Truth About Our Libertarian Age: Why the Dogma of Democracy Doesn't Always Make the World Better," *New Republic*, June 17, 2014, http://www.newrepublic.com/article/118043/our-libertarian-age-dogma-democracy-dogma-decline.

112 "libertarian, going to replace all other currencies": Jed McCaleb to BTCF, May 16, 2011.

114 MyBitcoin users went to the FBI's cybercrime unit: Adrianne Jeffries, "MyBitcoin.com Is Back: A Week After Vanishing with at Least $250 K. Worth of BTC, Site Claims It Was Hacked," BetaBeat, *New York Observer*, August 5, 2011, http://observer.com/2011/08/mybitcoin-disappeared-with-bitcoins/.

CHAPTER 11

115 "Have you ever thought about doing": Richard Bates, RUTT, January 22, 2015.

115 "I'm sure the authorities would be very interested": Richard Bates, RUTT, January 22, 2015.

116 He lied to Richard as one part of his effort to cover his tracks: RUTE GX 226D.

116 the site had generated $30,000 in commissions: RUTE GX 250.

116 in September Ross hired his first staff member: RUTE GX 250 and GX 240B.

117 he sold his pickup truck and moved to Sydney, Australia: David Kushner, "Dead End on Silk Road: Internet Crime Kingpin Ross Ulbricht's Big Fall," *Rolling Stone*, February 4, 2014, http://www.rollingstone.com/culture/news/dead-end-on-silk-road-internet-crime-kingpin-ross-ulbrichts-big-fall-20140204.

117 He would fit in his work around trips to Bondi beach: RUTE GX 240C.

118 "the biggest and strongest willed character I had met": RUTE GX 240B.

118 Variety Jones came up with a clever idea: RUTE GX 226D.

119 vendors in at least eleven countries: Nicolas Christin, "Traveling the Silk Road: A Measurement Analysis of a Large Anonymous Online Marketplace," Working Paper, November 28, 2012.

120 An academic study of Silk Road: Ibid.

120 In March, that amounted to nearly $90,000: RUTE GX 250.

121 In real life, DigitalInk's name was Jacob George: Ian Duncan, "Silk Road Drug Dealer Pleads Guilty," *Baltimore Sun*, November 5, 2013, http://articles.baltimoresun.com/2013-11-05/news/bs-md-silk-road -plea-20131105_1_drug-dealer-ross-william-ulbricht-jacob-theodore -george-iv.

CHAPTER 12

130 "He has not broken any rules and silk road": Sealed complaint against Charlie Shrem filed by IRS Special Agent Gary Alford, January 24, 2014.

132 Federal Reserve had held a daylong conference: Information about the conference is available at http://www.kc.frb.org/publications/research/ pscp/pscp-2012.cfm.

133 Canadian government announced the launch: Emily Jackson, "Royal Canadian Mint to Create Digital Currency," *Toronto Star*, April 11, 2012, http://www.thestar.com/business/2012/04/11/royal_canadian_ mint_to_create_digital_currency.html.

CHAPTER 13

137 "it funds a decent percentage of the overall": Sealed complaint against Charlie Shrem filed by IRS Special Agent Gary Alford, January 24, 2014.

141 group agreed that the bylaws for the foundation would be posted on GitHub: The bylaws are available at https://github.com/pmlaw/ The-Bitcoin-Foundation-Legal-Repo/tree/master/Bylaws.

CHAPTER 15

154 the company made $750 million for its investors: Eric Markowitz, "The $750 Million 'Mistake,'" *Inc.*, December 14, 2011, http://www .inc.com/articles/201112/argentine-entrepreneur-750-million-mistake .html.

158 the Argentinian government ordered his company, PayPal: "Paypal Suspends Domestic Transactions in Argentina," BBC News, September 17, 2012, http://www.bbc.com/news/technology-19605499.

159 35 percent lower than the rate available on the street: Historical data on the two different exchange rates available at http://dolarblue.net/historico/.

160 the first-ever Bitcoin Meetup in Argentina: Information on the meetups is available at http://www.meetup.com/bitcoin-Argentina/.

CHAPTER 16

167 Some $1.2 million worth of Bitcoin: Nicolas Christin, "Traveling the Silk Road: A Measurement Analysis of a Large Anonymous Online Marketplace," Working Paper, November 28, 2012.

167 seventy thousand different topics on Silk Road's forum: Eileen Ormsby, *Silk Road* (Sydney: Pan Macmillan Australia, 2014).

168 His work on Silk Road was done at an Internet café around the corner: Sealed complaint against Ross Ulbricht filed by FBI Special Agent Christopher Tarbell, September 27, 2013.

168 Over the summer, a Silk Road user had managed to follow a series of transactions: Ormsby.

169 paying the attacker $25,000: RUTE GX 250.

169 Ross explained that he was changing his writing style: Ormsby.

169 In November, Ross flew to Dominica: RUTE GX 291.

169 "put yourself in the shoes of a prosecutor": RUTE GX 225B.

170 Ross decided to help nob sell his kilogram: Superseding indictment against Ross Ulbricht filed by the Grand Jury for the District of Maryland, October 1, 2013.

171 Ross had always been somewhat skeptical: RUTE GX240B.

171 "beat up, then forced to send the Bitcoins he stole back": Superseding indictment against Ross Ulbricht filed by the Grand Jury for the District of Maryland, October 1, 2013. Ross has not yet been tried on the charges in the Maryland indictment and has not been found guilty on any counts related to murder.

CHAPTER 18

186 "PayPal will give citizens worldwide more": Eric Jackson, *PayPal Wars* (Washington, DC: WND Books, 2004).

187 Thiel advocating for floating structures: "Peter Thiel Offers $100,000 in Matching Donations to TSI, Makes Grant of $250,000," Seasteading Institute, February 10, 2010, http://www.seasteading.org/2010/02/peter-thiel-offers-100000-matching-donations-tsi-makes-grant-250000/.

187 aiming for the colonization of Mars: Adam Mann, "Elon Musk Wants to Build 80,000-Person Mars Colony," *Wired*, November 26, 2012, http://www.wired.com/2012/11/elon-musk-mars-colony/.

CHAPTER 19

190 In June 2012 the founders announced: BFL (Butterfly Labs) to BTCF, June 16, 2012.

190 a young Chinese immigrant in New York, Yifu Guo, announced: ngzhang to BTCF, September 17, 2012.

191 that power doubled again in just one month after Yifu's machines: Historical data on the hashing power available at https://blockchain.info/charts/hash-rate.

195 "This is a dark day for Bitcoin": "Breaking: The Blockchain Has Forked," *Bitcoin Trader*, March 11, 2013, http://www.thebitcointrader .com/2013/03/breaking-blockchain-has-forked.html.

196 "clarify the applicability of the regulations implementing": The FinCen guidance is available at http://fincen.gov/statutes_regs/guidance/html/ FIN-2013-G001.html.

CHAPTER 20

199 Martti Malmi posted an entry on his company's website: Martti Malmi, "SC5'er Intro: The Bitcoin Guy," SC5 website, February 5, 2013, http:// sc5.io/posts/sc5er-intro-the-bitcoin-guy.

205 "As a VC, my interest in the *Bitcoin* ecosystem is not ideological": Jeremy Liew, "Why VCs Love the Bitcoin Market," *TechCrunch*, April 5, 2013, http://techcrunch.com/2013/04/05/why-do-vcs-care-about-bitcoin/.

206 The BitInstant engineers congregated with their laptops: The scene in the office was captured in unreleased footage from Nicholas Mross's documentary *The Rise and Rise of Bitcoin* (2014), shared with the author.

207 Mark Karpeles assured his users that the problems were due to the volume of trade: Vitalik Buterin, "The Bitcoin Crash: An Examination," *Bitcoin Magazine*, April 13, 2013, https://bitcoinmagazine.com/4113/ the-bitcoin-crash-an-examination/.

CHAPTER 21

211 "For the time being, Bitcoin is in many ways": Felix Salmon, "The Bitcoin Bubble and the Future of Currency," *News Genius*, http://genius.com/ Felix-salmon-the-bitcoin-bubble-and-the-future-of-currency-annotated.

211 finally went public in the *New York Times*: Nathaniel Popper and Peter Lattman, "Never Mind Facebook; Winklevoss Twins Rule in Digital Money," *New York Times*, April 11, 2013, http://dealbook .nytimes.com/2013/04/11/as-big-investors-emerge-bitcoin-gets-ready -for-its-close-up/?_r=0.

211 a national television station in China broadcast a half-hour segment: The May 3, 2013, segment is available at http://jingji.cntv.cn/2013/05/03/ VIDE1367596319388137.shtml.

211 $2 million into BitPay: The announcement is available at http://www .marketwatch.com/story/bitpay-raises-2-million-led-by-founders-fund -2013-05-16.

212 $5 million into Coinbase: The announcement is available at https:// www.usv.com/post/coinbase.

213 Mark was sued in a Seattle court by CoinLab: Complaint filed by Coin-Lab against Mt. Gox on May 2, 2013, in United States District Court for the Western District of Washington.

213 money in Mt. Gox's two American bank accounts—some $5 million— was seized: Romain Dillet, "Feds Seize Another $2.1 Million from Mt. Gox, Adding Up to $5 Million," *TechCrunch*, August 23, 2013, http://techcrunch.com/2013/08/23/feds-seize-another-2-1-million -from-mt-gox-adding-up-to-5-million/.

CHAPTER 22

218 federal prosecutors arrested the operators of Liberty Reserve: Information on the arrest is available at http://www.justice.gov/usao/nys/press releases/May13/LibertyReservePR.php.

218 the top financial regulator in California sent the Bitcoin Foundation: The letter was posted by the executive director of the foundation at http:// www.forbes.com/sites/jonmatonis/2013/06/23/bitcoin-foundation -receives-cease-and-desist-order-from-california/.

224 announced a few days after Charlie shut down BitInstant: Erik Voorhees to BTCF, July 17, 2013.

225 one-millionth registered account: Eileen Ormsby, *Silk Road* (Sydney: Pan Macmillan Australia, 2014).

225 commissions collected by the site often approached over $10,000 a day: RUTE GX 250.

225 Ross agreed to pay $100,000 up front: RUTE GX 241.

226 "Don't want to be a pain here": Sealed complaint against Ross Ulbricht filed by FBI Special Agent Christopher Tarbell, September 27, 2013.

226 paid for with 3,000 Bitcoins, or roughly $500,000: Letter opposing Ross Ulbricht's release on bail, filed by Assistant United States Attorney Serrin Turner, November 20, 2013. These alleged murders and the chats between Ross and redandwhite were discussed during Ross Ulbricht's trial, but Ross was not charged with any counts of murder for hire and Canadian police never found any evidence of any suspicious deaths during this time that might be tied to Ross.

227 He moved out of his friend's apartment in June: Sealed complaint against Ross Ulbricht filed by FBI Special Agent Christopher Tarbell, September 27, 2013.

227 "encrypt and backup important files": Letter opposing Ross Ulbricht's release on bail, filed by Assistant United States Attorney Serrin Turner, November 20, 2013.

228 "Without going into details, the stress of being": Dread Pirate Roberts to Silk Road forum, September 20, 2013.

228 Ross assigned Variety Jones: RUTE GX 241.

228 When agents knocked on the door: Sealed complaint against Ross Ulbricht filed by FBI Special Agent Christopher Tarbell, September 27, 2013.

229 Ross changed apartments: Thomas Kiernan, RUTT, January 22, 2013.

CHAPTER 23

238 opened 350,000 free Blockchain.info wallets: Data on wallets available at https://blockchain.info/charts/my-wallet-n-users.

240 At a Bitcoin Meetup in July 2013, two hundred: Information on the meetups is available at http://www.meetup.com/bitcoin-Argentina/.

241 "You don't have to be battling": Jose Crettaz, "Bitcoin: Fiebre argentina por la máquina de dinero digital," *La Nación*, June 30, 2013, http://www.lanacion.com.ar/1596773-bitcoin-pasion-argentina-por-la-nueva-maquina-de-hacer-billetes-digitales.

241 the peso was down some 25 percent: Historical data on the two different exchange rates available at http://dolarblue.net/historico/.

CHAPTER 24

245 wobbling out of control in late September: All details in this paragraph are from RUTE GS 241.

245 "I have poison oak rash": RUTE GX 325.

246 The next day he spent the morning working: Jared Der-Yeghiayan, RUTT, January 14, 2015.

246 He headed to the far side of the library: RUTE GX 128H.

246 "sure, someone could stand behind you": RUTE GX 225B.

247 "dread: im ok, you?": RUTE GX 129C.

247 There were 25,689 orders in transit: Numbers are taken from a screen shot of Ross's computer on the day of his arrest; it was submitted by the government as evidence before Ross's trial.

247 This was the signal that cirrus had: Jared Der-Yeghiayan, RUTT, January 14, 2015.

248 "I'm so sick of you," the woman shouted: David Kushner, "Dead End on Silk Road: Internet Crime Kingpin Ross Ulbricht's Big Fall," *Rolling Stone*, February 4, 2014, http://www.rollingstone.com/culture/news/dead-end-on-silk-road-internet-crime-kingpin-ross-ulbrichts-big-fall-20140204.

248 As Ross turned around to see what was: Thomas Kiernan, RUTT, January 22, 2013.

248 did so by searching on Google through old: Gary Alford, RUTT, January 26, 2013.

249 Users of Silk Road visiting the hidden site that morning: "FBI Arrests Silk Road Drug Site Suspect," BBC News, October 2, 2013, http://www.bbc.com/news/technology-24373759.

251 In court, Ross was in shackles: "Attorney Denies Charges That San Francisco Man Operated Encrypted Drug Website," Associated Press, October 4, 2013.

CHAPTER 25

257 China's previous experience with a successful virtual currency: Mark Lee, "China Bans Online Virtual Money Dealing for Minors," Bloomberg, June 22, 2010, http://www.bloomberg.com/news/articles/2010-06-22/tencent-shares-fall-after-china-announces-virtual-currency-ban-for-minors.

259 The reporter for Channel 2 tracked: The May 3, 2013, segment is available at http://jingji.cntv.cn/2013/05/03/VIDE1367596319388137.shtml.

260 Macao, seven times bigger, in revenue terms, than Las Vegas: Charles Riley, "Macau's Gambling Industry Dwarfs Vegas," CNNMoney, January 6, 2014, http://money.cnn.com/2014/01/06/news/macau-casino-gambling/index.html.

261 a division of Baidu, the search engine giant and the fifth-most-visited website in the world, announced: Vitalik Buterin, "Baidu Jiasule and the Chinese Bitcoin Community," *Bitcoin Magazine*, October 16,

2013, https://bitcoinmagazine.com/7492/baidu-jiasule-and-the-chinese-bitcoin-community/.

262 John Donahoe, said in an interview: Andrea Felsted, "Ebay to Expand the Range of Digital Currencies It Accepts," *Financial Times*, November 3, 2013.

CHAPTER 26

266 "long-term promise, particularly if the innovations": Ben Bernanke letter to Senate Committee on Homeland Security and Governmental Affairs, September 6, 2013.

268 A story the previous week in Xinhua: Xinhua story is available at http://news.xinhuanet.com/fortune/2013-11/15/c_118148623.htm.

269 "I do not want to shut down or stamp out Bitcoin": Morgan Peck, "If Senators Really Like Bitcoin They Should Encourage Banks to Cooperate," *IEEE Spectrum*, November 21, 2014, http://spectrum.ieee.org/tech-talk/computing/networks/us-senate-.

269 Silk Road 2.0 showed up on the dark web: Eileen Ormsby, "Remember, Remember . . . Silk Road Redux," *All Things Vice*, November 7, 2013, http://allthingsvice.com/2013/11/07/remember-remember-silk-road-redux/.

270 The number of Blockchain.info wallets: Data on wallets available at https://blockchain.info/charts/my-wallet-n-users.

271 But the relatively apathetic public response: David Lauter, "Public Largely Tunes Out NSA Surveillance Debate, Poll Finds," *Los Angeles Times*, January 20, 2014.

271 "We see the intrinsic value of Bitcoin": Gil Luria, "Bitcoin: Intrinsic Value as Conduit for Disruptive Payment Network Technology," Wedbush Equity Research, December 1, 2013.

272 "emerge as a serious competitor": David Woo, "Bitcoin: A First Assessment," Bank of America Merrill Lynch FX and Rates Research, December 5, 2013.

274 The good news was that the agencies: The Chinese government statement is available at http://www.pbc.gov.cn/publish/goutongjiaoliu/524/2013/20131205153156832222251/20131 205153156832222251_.html.

CHAPTER 27

286 Krugman focused largely on Bitcoin's claim: Paul Krugman, "Bitcoin Is Evil," *New York Times*, December 28, 2013, http://krugman.blogs.nytimes.com/2013/12/28/bitcoin-is-evil/.

286 Cowen, meanwhile, argued: Tyler Cowen, "How and Why Bitcoin Will Plummet in Price," *Marginal Revolution*, December 30, 2013, http://marginalrevolution.com/marginalrevolution/2013/12/how-and-why-bitcoin-will-plummet-in-price.html.

287 "to an extent that makes a sub-Saharan African kleptocracy": Charles Stross, "Why I Want Bitcoin to Die in a Fire," *Charlie's Diary*, December 18, 2013, http://www.antipope.org/charlie/blog-static/2013/12/why-i-want-bitcoin-to-die-in-a.html.

289 "It represents a remarkable conceptual": Francois Velde, "Bitcoin: A Primer," *Chicago Fed Letter*, December 2013.

289 Overstock announced that it would begin: The announcement is available at http://blog.coinbase.com/post/72787431702/coinbase-and-overstock-com-announce-largest.

290 Overstock processed more than $100,000 in orders: Sales data available at http://www.prweb.com/releases/bitcoin2014Keynote/PatrickByrne/prweb 11699797.htm.

CHAPTER 28

291 Thiel called him the "firefighter-in-chief": Evelyn M. Rusli, "A King of Connections Is Tech's Go-To Guy," *New York Times*, November 5, 2011, http://www.nytimes.com/2011/11/06/business/reid-hoffman-of-linkedin-has-become-the-go-to-guy-of-tech.html?pagewanted=all.

291 Hoffman later introduced Thiel to Mark Zuckerberg: David Kirkpatrick, *The Facebook Effect* (New York: Simon & Schuster, 2010).

294 "The gulf between what the press and many": Marc Andreessen, "Why Bitcoin Matters," DealBook, *New York Times*, January 21, 2014, http://dealbook.nytimes.com/2014/01/21/why-bitcoin-matters/.

295 He believed that it could help open the door: A transcript of Balaji's talk at Startup School 2013 is available at https://nydwracu.word press.com/2013/10/28/transcript-balaji-srinivasan-on-silicon-valleys-ultimate-exit/.

299 The prosecutors had e-mails in which: Sealed complaint against Charlie Shrem filed by IRS Special Agent Gary Alford, January 24, 2014.

300 "If you want to develop a virtual currency": The press release announcing Charlie's arrest is available at http://www.justice.gov/usao/nys/pressreleases/January14/SchremFaiellaChargesPR.php.

303 told CNBC in late January: Jamie Dimon, interviewed on CNBC, January 23, 2014.

CHAPTER 29

309 In a statement, Mark explained: While material from the Mt. Gox website has been deleted, the full statement is still available at http://pando
.com/2014/02/10/blame-game-embattled-mt-gox-points-to-flaws-in
-bitcoin-protocol-bitcoin-community-calls-bs/.

310 He was wearing a short-sleeved shirt: The confrontation was recorded
and is viewable at https://www.youtube.com/watch?v=ob9Ak1t09Ao.

315 "This tragic violation of the trust of users": The statement is available at http://blog.coinbase.com/post/77766809700/joint-statement
-regarding-mtgox.

316 lawyers in Chicago and Denver filed a lawsuit: Jonathan Stempel and
Emily Flitter, "Mt. Gox Sued in United States over Bitcoin Losses," Reuters, February 28, 2014, http://www.reuters.com/article/2014/02/28/
bitcoin-mtgox-lawsuit-idUSL1N0LX1QK20140228.

317 An academic study in 2013: Tyler Moore and Nicolas Christin. "Beware
the Middleman: Empirical Analysis of Bitcoin-Exchange Risk." In
Ahmad-Reza Sadeghi, editor, *Financial Cryptography*, volume 7859 of
Lecture Notes in Computer Science (New York: Springer, 2013).

317 "The only way to stabilise the system is": Izabella Kaminska, "Magic:
The Undercapitalized Gathering Online," *FT Alphaville*, March 3,
2014, http://ftalphaville.ft.com/2014/03/03/1787992/magic-the-under
capitalised-gathering-online/.

CHAPTER 30

319 The *Newsweek* reporter, Leah McGrath Goodman, had: Leah McGrath
Goodman, "The Face Behind Bitcoin," *Newsweek*, March 6, 2014, http://
www.newsweek.com/2014/03/14/face-behind-bitcoin-247957.html.

321 "Why did you create Bitcoin, sir?": The video of Dorian Nakamoto leaving his house is viewable at http://www.theguardian.com/
technology/2014/mar/07/satoshi-nakamoto-denies-inventing-bitcoin.

323 "simply be an old man saying ANYTHING": Gavin's letter to McGrath
Goodman is available at http://www.reddit.com/r/bitcoin/comments/
1zqjq6/open_letter_to_leah_mcgrath/.

323 In an Amazon review of Danish butter cookies: The review is available
at http://www.amazon.com/review/R3U92F9YRUSF37.

323 The AP's story and video from its interview: The interview is viewable at https://www.youtube.com/watch?x-yt-ts=1422579428&x-yt-cl=
85114404&v=GrrtA6IoR_E.

324 An Argentinian security expert, Sergio Lerner, had done: Sergio Demian
Lerner, "TheWellDeservedFortuneofSatoshiNakamoto,BitcoinCreator,

Visionary and Genius," Bitslog, April 17, 2013, https://bitslog.wordpress
.com/2013/04/17/the-well-deserved-fortune-of-satoshi-nakamoto/.

333 "Friends, citizens, Bitcoiners, there is nothing": Charlie's speech is
viewable at https://www.youtube.com/watch?v=xH7mCO5EnDU.

334 "I think it's very obvious to all of us": Gregory Ferenstein, "Google's
Jared Cohen: It's 'Obvious' Bitcoin-Like Currencies Are 'Inevitable,'"
TechCrunch, March 8, 2014, http://techcrunch.com/2014/03/08/
googles-jared-cohen-its-obvious-bitcoin-like-currencies-are-inevitable/.

335 "You don't get the new technology from": Andreessen's comments are
from his speech at Coinsummit 2014, which is viewable at https://www
.youtube.com/watch?v=iir5J6Z3Z1Q.

CHAPTER 31

339 Nick's writing: Nick's writings are available at http://unenumerated
.blogspot.com/.

339-40 Most bizarrely, Nick altered the dates: the dates that Nick later put
on the posts are at the top of each post. But the URL addresses of the
posts still show the original posting date. For instance, his post on "Bit
Gold Markets" says that it was written on December 27, 2008, but the
URL is http://unenumerated.blogspot.com/2008/04/bit-gold-markets
.html#links.

339 "repeated use of 'of course' without isolating commas": Skye Grey, "Satoshi Nakamoto Is (Probably) Nick Szabo," *LikeinaMirror*, December 1,
2013, https://likeinamirror.wordpress.com/2013/12/01/satoshi-nakamoto
-is-probably-nick-szabo/.

348 a hacker demanding ransom was targeting Hal: Robert McMillan, "An
Extortionist Has Been Making Life Hell for Bitcoin's Earliest Adopters," *Wired*, December 29, 2014, http://www.wired.com/2014/12/finney
-swat/.

353 The United States Marshals Service had auctioned off the 29,655: Tim
Draper's announcement is available at https://medium.com/mirror-blog/
tim-draper-wins-govt-auction-partners-with-vaurum-to-provide-bit
coin-liquidity-in-emerging-markets-88f04a1d8598.

353 Wences officially announced the $20 million: The Xapo announcement
is available at https://blog.xapo.com/xapo-raises-20-million-investment
-led-by-greylock-partne/.

354 Gates had initially bet against the open Internet and built a closed network: Kathy Rebello, "Inside Microsoft: The Untold Story of How the Internet Forced Bill Gates to Reverse Course," *BusinessWeek*, July 15, 1996.

INDEX

Alcor Life Extension Foundation, 7

Alibaba (Chinese Internet company), 261

Alice (hypothetical user), 9, 11, 21–23, 358–359

Alipay (Chinese payment processor), 260–261

Allen & Co., 181, 292, 349, 353

altoid (screen name), 69, 248. *See also* Ulbricht, Ross

Andreessen Horowitz, 186, 192, 329

Andreessen, Marc, 181, 186–187, 293–295, 303, 335

Andresen, Gavin
  beginnings with Bitcoin, 44–47, 49–50, 323
  as Bitcoin central figure, 59–62
  Bitcoin mining, 53, 192–197, 329
  Bitcoin promotion, 75–76, 101–106
  creation of Bitcoin Foundation, 138, 141–142
  dealing with scandals, 99
  relationship with Satoshi, 55–56, 80–86
  responding to Mt. Gox collapse, 309

2014 Bitcoin Pacifica (Lake Tahoe), 346–348

Anoncoin (digital currency), 270–271

Anti-state.com (website), 29

Argentina, 153–161, 182, 240–242, 259, 277–280, 286, 349–353

ASIC (computer chip), 189–190, 259, 329–330

Assange, Julian, 56–57

Athey, Susan, 345

Atlantis, 245

Australia, 44–45, 117, 168, 171

Automated Clearing House (ACH), 133

Avalon (ASIC), 190, 206

Back, Adam, 17–22, 339, 348

Baidu (Chinese search engine), 261–262

bank bailout of 2008, 32, 111

Bank of America, 272

Bates, Richard, 75–77, 115–116

Bedier, Osama, 101

bee-te-bee (Chinese Bitcoin), 255–256

Beijing Summer Olympics (2008), 145

Benchmark Capital, 282, 293, 305

Bernanke, Ben, 266

Bezos, Jeff, 353

Bharara, Preet, 299–300

Bill and Melinda Gates Foundation, 353–355

Bitcoin
  arrival of Gavin Andresen, 44–47
  arrival of Martti Malmi, 29–30
  building trust, 24–25, 33, 48, 61–62, 69, 99–100, 279, 315, 339
  buying/selling with, 43–44, 82, 119–120, 129–130
  changing business model, 236–239
  characterization as "cryptocurrency," 36
  comparison to gold, 157–158, 165, 182
  comparison to paper money, 219, 286–287
  creation and operation of original code, 4–6, 20–24
  disappearance of Satoshi Nakamoto, x–xiii
  hacking and scandals, 91–99
  increasing price/value, 38, 66–69, 79, 81–85, 89–91, 131, 175, 180, 184, 193–196, 204–206, 210–211, 250–253, 262–264, 267, 271–275, 284–285
  legality/government regulation, 196–198, 251
  limitations based upon computers, 347
  Mt. Gox collapse busts bubble, 308–317
  origin and ideology, vii–xv, 5, 113–114
  as Ponzi scheme, 220
  proof-of-work, 18–19
Bitcoin Foundation
  candidacy of Bobby Lee, 345

dealing with Bitcoin collapse, 314–315
  Gavin Andresen as member, 192
  involvement in Senate hearing, 265–267, 270, 300–302
  Patrick Murck as member, 176
  planning and creation of, 138–142
  regulatory problems, 217–219, 233–236
  resignation of Charlie Shrem, 302
Bitcoinica, 237
Bitcoin Investment Trust, 314
Bitcoin Meetups. *See* conferences (Bitcoin and others)
Bitcoin mining
  about process vulnerability, 41–42
  creating blocks and recording transactions, 359–361
  creation of ASIC chip, 189–192, 259, 329–330
  creation of Avalon chip, 190, 206
  formation of mining companies, 294–295, 328–329
  formation of mining pools, 192–194
  GPU technology, 42, 56, 189–191
  growth in China, 259–261, 329
  Litecoin mining, 283
  more users increased difficulty, 53
  role in securing system, 100
  Satoshi Nakamoto patterns, 324
  specialized computers/computing power, 105, 170, 190, 233, 324, 330, 347
The Bitcoin Show (TV program), 102, 128
Bitcoin software
  about operation, 23, 357–362
  beta testing, 25–26, 58
  changes to code, 22–24, 35–39, 43–46, 55–58, 61–62, 141, 309, 346–347
  creating/maintaining protocol, x, 5–6, 32, 99, 215–216

creation and launch, xiv, 30–31, 319, 346

downloads, 49–51, 80, 237, 261

Google interest, 100–102

hard fork, 193, 195

"1 RETURN" bug, 56

role of public-key cryptography, 9–10, 17–18

running on Macintosh, 41

transaction malleability problem, 309–314

updates and old versions, 37, 59, 193–195

version 0.2, 37

version 0.3, 47–48

version 0.319, 59

version 0.7, 194–195

version 0.8, 194–195

The Bitcoin Trader (blog), 195

Bitcoin White Paper, 21, 45, 339

Bitfury, 330

bit gold, 18, 338–339

BitInstant. *See also* Shrem, Charlie

attracting investors, 130–135

creation and function, 128–130

dealing with problems and competitors, 201–207

hacker penetration, 150

investment by David Azar, 134, 150–151

investment by Roger Ver, 128

investment by Winklevoss twins, ix, 173–176, 211–215

involvement of Erik Voorhees, 135–137

management problems, 220–222

regulatory problems, 222–224

trading volume, 201, 205–207

BitLicense, 302, 317

Bitomat (Polish exchange), 97–98

BitPagos (Argentinian payment service), 278–279

BitPay, 134, 211, 219, 272

Bitstamp (Slovenian exchange) about founding, 203

attendance at 2014 Bitcoin Pacifica, 252–253, 337

regulation of virtual currencies, 271

response to Mt. Gox collapse, 309–310, 315

surpassing Mt. Gox volume, 236

trading volume, 262–263, 267

working with banks, 327

blind digital signatures, 12

blockchain

banking interest in the technology, 324–328

Bitcoin transfers, 97–98, 133, 148, 182, 203–204, 235–237

creation and function, 21–26, 43, 55, 61, 340

dealing with hard fork, 193–194

generating new coins, 361–362

increasing file size, 100–101

sidechains, 348

use by mining pools, 191–194

use by money transfer projects, 188–189, 336

winning acceptance and approval, 269–274, 289–290, 345

winning blocks, 361

Blockchain.info, 237–241, 252, 270, 315, 330–331

Blodget, Henry, 182–184

Bloomberg, Michael, 144, 325

b-money, 339

Branson, Richard, 297

Briger, Pete, 163–165, 201, 236, 252–253, 281–283, 287–288, 302, 342–343. *See also* Fortress Investment Group

Brito, Jerry, 79–80

Bruno, Joe Bel, 322

BTC China, 255–264, 267–269, 275, 284–285, 300, 315, 343–345. *See also* China

BTC Guild, 195

BTC King (screen name). *See* Faiella, Robert

Buffett, Warren, 353
Burges, Kolin, 310–312
Business Insider, 184
BusinessWeek, 197
Byrne, Patrick, 289

Canada, launch of Mint Chip, 133
Carper, Thomas (senator), 235,
    267–268
Cary, Nic, 239, 252, 296–298,
    333
Casares, Belle, 154, 162, 243, 352
Casares, Wences. *See also* Lemon
    Digital Wallet and Xapo
    background and arrival at Bitcoin,
        153–165
    Bitcoin as commodity, 274
    Bitcoin holdings, 287–288
    Bitcoin promotion, 179–180,
        185–187, 197, 209–210
    Bitcoin promotion in Argentina,
        240–242
    conference attendance, 181–185,
        214–216, 349, 351–355
    development of Lemon Digital
        Wallet, 201–205
    sale of Lemon Digital Wallet, 252,
        280–283
    seeking business investors,
        291–296
    startup business financing,
        305–306
    2013 Argentina, Bitcoin meeting,
        277
    Xapo founding and operations,
        349–351
Casascius coins, 126–127
chronicpain (screen name). *See*
    Green, Curtis
cimon (screen name). *See* Variety
    Jones [vj]
cirrus (screen name), 246–248
Chaum, David, 10, 12, 23, 71. *See
    also* DigiCash
China, xiii, 128, 183, 190–191,

273–275, 280, 329. *See also* BTC
    China
CIA. *See* U.S. Central Intelligence
    Agency
Coinapult, 174, 338
Coinbase (Bitcoin service). *See also*
    Ehrsam, Fred
    about the founding and operation,
        203–204, 211–213
    investment by Andreessen
        Horowitz, 293–295
    maintaining private keys, 281
    regulation of virtual currencies,
        271
    regulatory compliance, 236–237
    response to Mt. Gox collapse, 315
    transaction fees, 290
    working with banks, 305–306
CoinLab, 138, 144, 200, 213
COIN (Nasdaq ticker symbol), 353
Collins, John, 265–266
conferences (Bitcoin and others)
    2011 CIA interest in Bitcoin, 81
    2011 NYC Bitcoin World Expo,
        102–106, 135
    2011 Thailand, Bitcoin, 104
    2012 Amsterdam, Bitcoin, 104,
        297–298
    2012 Federal Reserve on money
        transfer, 132–133
    2012 NYC, Bitcoin, 104
    2013 Allen & Co., 181, 349
    2013 Argentina, Bitcoin, 277–283
    2013 San Jose, Bitcoin, 214–216
    2014 Allen & Co., 262, 349,
        353–355
    2014 Austin, Bitcoin, 331–336
    2014 Bitcoin Pacifica (Lake
        Tahoe), 337–345
    2014 SXSW, 334–336
    2014 Utrecht technology, 298
*The Construction and Operation of
    Clandestine Drug Laboratories*
    (Jack B. Nimble), 69
Costollo, Dick, 181

Cowen, Tyler, 286
CRASH (CRypto caSH), 12
credit cards
  Bitcoin as replacement, 23,
    158–160, 235, 292
  digital wallets and, 101, 154, 209
  disputes and chargebacks, 64, 134,
    343
  lack of privacy, 11
  Target Corporation, data breach,
    288–289
  transaction fees, xii, 102, 240–
    241, 272, 277–278, 290, 343
  WikiLeaks blockade, 57
Crisis Strategy Draft, 313–315
cryogenics, 7
cryptocurrency, 36
Cryptonomicon (Stephenson), 19, 252
currency debasement, 30–31
Cypherpunk Manifesto, 8–12
Cypherpunks
  awareness of privacy and data
    vulnerability, 8–9
  conceptualizing future of money,
    11–13, 16
  facing digital money obstacles,
    19–20
  philosophical influences, 70
  termination of mailing list, 20

Dai, Wei, 19–20
Darkcoin (digital currency),
  270–271
Debt: The First 5,000 Years
  (Graeber), 157, 179
decentralized systems/technology.
  See also Blockchain
  about Bitcoin ideology, 236
  Bitcoin comparison to gold, x
  building Bitcoin system, 55–56,
    141, 292–294
  development of payment systems,
    129, 133
  disadvantages of centralization,
    113–114

Internet as, 182
Occupy Wall Street movement
  and, 111
open source software and, 45–47
P2P Foundation and, 30
regulatory compliance, 269–270
resolving problems, 195
Silk Road and, 118
trend toward centralization,
  99–100, 347
Der-Yeghiayan, Jared, 246–248
DigiCash, 12, 19, 21, 23, 26, 158.
  See also Chaum, David
digital currency
  Anoncoin, 270–271
  Chinese potential, 260–261
  creation of early systems, 12–13
  Darkcoin, 271
  Finney experimentation, 5
  Greenspan prediction, 17
  Liberty Reserve, 218
  Mint Chip, 133
  Q coin, 257, 260–261, 268
DigitalInk (screen name). See
  George, Jacob
Dimon, Jamie, 303–306
Dixon, Chris, 181–182, 186, 294
Dodd, Nigel, 16
Donahoe, John, 262
Donald, James, 24
Draper, Tim, 353
Dread Pirate Roberts [DPR] (screen
  name), 118, 121, 168–169, 171,
  213, 225, 227, 248. See also
  Ulbricht, Ross
drugs/drug trafficking. See Silk
  Road

eDonkey (file sharing website),
  50–51
Ehrsam, Fred, 334–336. See also
  Coinbase (Bitcoin service)
Electronic Frontier Foundation, 80,
  270
Eleuthria (screen name), 195

encryption technology, 8–12. *See also* Public-key cryptography

exchange-traded funds (ETF), 222, 250

Extropians, 11

Facebook, 145

Faiella, Robert (aka BTC King), 130, 138, 299

The Far Wilds (online game), 50–51

FBI. *See* U.S. Federal Bureau of Investigation

Federal Reserve. *See* U.S. Federal Reserve

Financial Crimes Enforcement Unit [FinCen] (Treasury Department), 138, 196–197, 201, 234–235, 266, 325

Financial Times, 262, 317

Finney, Fran, 3

Finney, Hal
defense of Bitcoin system, 24–27
introduction to Bitcoin, 3–8
Lou Gehrig's disease diagnosis, 27
return to Bitcoin community, 59–60
role in PGP, 10, 13

Finney, Jason, 27

FirstMark Capital, 144, 147–149, 176

Forbes, 80, 96

Fortress Investment Group, 180, 217–219, 252, 272–273. *See also* Briger, Pete

Founders Fund, 187, 211

4chan (hacker message board), 75

Freeman, Ian, 75–76

Free State Project, 107–110

Free Talk Live (radio program), 75–78, 108

Freis, James, 325

FriendlyChemist (screen name), 225–226

Gandalf (computer chip), 329

Garzik, Jeff, 83–84, 92, 99, 190, 196, 348

Gates, Bill, 353–355, 385n

Gawker (website), 83–84

George, Jacob (aka DigitalInk), 121

George Mason University, 80

Georgia, Republic of, 330

GitHub, 141

Goldman Sachs, 324–326

gold standard, x, 15–16, 31–32, 45, 109, 157–158

Gonzague, 312–315

Goodman, Leah McGrath, 319–324

Google, 101–103, 187, 248–249, 283, 304–305, 314–315, 334

Google Wallet, 101

government regulation/investigation
arrest of Roger Ver, 77–78
arrest of Ross Ulbricht, 170–171
BitInstant, 222–224
BTC China, 273–275
Erik Voorhees, 224–225
PGP and Zimmerman, 10
virtual currencies, 66–67, 196–198, 235

Graeber, David, 157

Great Depression, 31

Great Recession, banking crisis of 2008, 32, 111

Green, Curtis (aka chronicpain), 116, 170–171, 225, 249, 332

Greenspan, Alan, 17

hackers/hacking
Bitcoin vulnerability, xiv, 24, 154, 201, 215
BitInstant penetration, 150
message boards, 75
Mt. Gox penetration, 67–69, 82–83, 90–96, 99, 114, 205–207
ransom demands/payments, 82, 150, 169, 347–348
Silk Road penetration, 168–169, 225, 248–251
Target data breach, 288–289
vulnerability of private information, xii, 19

Hanecz, Laszlo, 41–44, 48, 58, 189, 215
Hashcash, 17–22, 339, 348
hash functions, 22, 25–26, 41–42, 136, 359–362
Hearn, Mike, 80–81, 99–100, 101–103, 320
Hoffman, Reid, 181, 183, 291–293, 353
Horowitz, Ben, 192, 335. *See also* Andreessen Horowitz
Huang Xiaoyu, 255–256, 258
Hughes, Eric, 8, 11–12
Huobi (Chinese exchange), 285

Iceland, 330
India, 350
  WikiLeaks blockade, 57
Internet relay chat (IRC), 41, 54, 67, 196
Internet terrorism. *See* hackers/hacking

Jack B. Nimble, 69
Jiasule (Chinese security service), 261
Johnston, David, 204
JP Morgan Chase & Co., 202, 218, 303–306, 327
Juno Moneta (Roman god), 17

Kaminska, Izabella, 317
Karpeles, Mark. *See also* Mt. Gox (Bitcoin exchange)
  arrival at Mt. Gox, 65–68
  becoming Mt. Gox owner, 68–69
  control over Mt. Gox code, 99
  lack of management skill, 127–128, 233, 307–309
  marginalization in Bitcoin future, 331
  NYC Bitcoin Expo 2011, 103–105
  struggling with Mt. Gox growth and problems, 140–141, 200–201
  vacating Bitcoin Foundation seat, 345

Kodric, Nejc, 253
Kraken (Bitcoin exchange), 315
Krugman, Paul, 286
Kutscher, Ashton, 335
Kuzmin, Alexander, 135–136

La Nación (newspaper), 241
Larsen, Chris, 325–326
Lawsky, Benjamin, 225, 300–302, 304, 317
Lee, Bobby, 256, 261–265, 267, 273, 300, 343–345
Lee, Charlie, 100, 103, 105, 256, 283
Lemon Digital Wallet, 154, 160–162, 179–180, 242–244, 252, 261, 280–283, 351. *See also* Casares, Wences
Lenihan, Larry, 144, 176
Lerner, Sergio, 324
Levchin, Max, 185–186, 349, 353
Levine, John, 24–25
Liberty Reserve (online currency), 218
Liew, Jeremy, 300–301
Lilla, Mark, 111
Ling Kang, 268, 274–275
LinkedIn (networking site), 291–292
Litecoin, 256, 283, 286
Luria, Gil, 271–272

MagicalTux (screen name). *See* Karpeles, Mark
Magic: The Gathering Online Exchange. See Mt. Gox (Bitcoin exchange)
Magic: The Gathering (online game), 51, 77
Maguire Ventures, 149
Makan, Divesh, 293
Malka, Micky, 179–180, 201–203, 210–212, 236, 242, 252, 282–283
Malmi, Martti

Malmi, Martti (*cont.*)
  beginning connection to Bitcoin,
    29–30, 200
  entry into Bitcoin operations,
    33–39, 44–50
  exchange services and forums,
    53–54
  making Silk Road work, 72, 84
  reduced role in Bitcoin, 58–60
  return to Bitcoin community, 348
  running Bitcoin website, 66,
    80–82, 86
Marcus, David, 158–159, 181, 184–
  185, 201, 216, 281, 292, 349
Mark Twain Bank, 12
Maxwell, Gregory, 348
McCaleb, Jed
  creation of Ripple, 187, 325
  exclusion from Bitcoin
    Foundation, 139
  founding of Mt. Gox, 50–53
  handling Mt. Gox disputes and
    problems, 63–65
  Mt. Gox account hacked, 90,
    94–95
  partnership with Mark Karpeles,
    65–69
  2011 NYC Bitcoin Expo, 103–105
  2014 Bitcoin Pacifica (Lake
    Tahoe), 337
  views on political ideology,
    112–113
meetups. *See* conferences (Bitcoin
  and others)
Memory Dealers, 78–79, 93, 126.
  See also Ver, Roger
Mercatus Center, 80
Miller, Ira, 137, 174, 176, 201
mining. *See* Bitcoin mining
Mint Chip (Canadian digital
  currency), 133
money/monetary systems
  about origins and determination of
    value, 15–17
  basis of market economy, 11

Bitcoin as form of, 82, 286
Bitcoin as replacement for gold,
  182
currency debasement, 30–31
evolution of credit, 157
gold standard, x, 15–16, 31–32,
  45, 109, 157–158
jokes on nature of, 146
money laundering
  Bank Secrecy Act enforcement,
    196
  Bitcoin as form of, 84, 303
  investigation and arrests, ix,
    224–225, 249, 266–267, 298
  PayPal investigation as, 186, 216
  regulatory compliance, 92, 203,
    218
Mongolia, 330
Morehead, Dan
  gathering for Lake Tahoe poker
    game, viii, xii–xiii
  investment in Slovenian exchange,
    236
  invited to work with Fortress, 217
  meeting for first Bitcoin Pacifica,
    251–253
  meeting for second Bitcoin
    Pacifica, 337–342
  relocation out of Fortress offices,
    342–343
Morton, Chris, 221–222
MS Haberdasher (screen name), 170
Mt. Gox (Bitcoin exchange)
  about the founding and growth,
    50–54
  arrival of Mark Karpeles, 65–69
  contending with competitors,
    97–102
  federal seizure of assets, 213–214
  government regulation, 66–67
  hacker vulnerability/penetration,
    67–68, 82–83, 89–96
  management problems, 307–309
  plans for closure, 312–314
  reaction to collapse and

bankruptcy, 314–316
shutting down exchange, 193, 205–207
trading disputes and problems, 63–65, 85–86, 199–207
trading volume, 53, 79, 201, 203, 207, 209–210
transaction malleability problem, 309–312
Murck, Patrick, 139–142, 176, 233–236, 265–267, 269, 302
Murdoch, James, 181
Murdoch, Rupert, 353
Murrone, Federico ("Fede"), 159–160, 201, 280–281, 349, 351
Musk, Elon, 187
MyBitcoin (online Bitcoin wallet), 98–99, 113–114, 237

Napster (music sharing service), 35, 50–51, 347
Nasdaq Stock Exchange, 222, 353
Nas (rapper), 335
National Security Agency (NSA), 8, 271
Nebseny, Val, 329–330
Nelson, Gareth, 129–130
NewLibertyStandard (screen name), 37–39, 41, 48, 69
Newsweek, 319–321, 324–325, 342, 347
New York, State of. *See* Government regulation/investigation
New York Times, 211, 293, 303, 365
"99 percenters." *See* Occupy Wall Street movement
Niven, Larry, 7
Nixon, Richard, 31
nob (screen name), 121, 170–171, 332
Novogratz, Mike, 180

O'Brien, Eric, 210

Occupy Wall Street movement, xi–xii, 111–112, 140, 157, 287, 331, 336
OKCoin, 344
Olympic Summer Games (Beijing, 2008), 145
"1 percenters," wealth distribution, 287, 336
Overstock (online retailer), 289–290
Ovitz, Michael, 183

Pantera Capital, 217, 343
Patagon, 162
Paul, Ron, 110–111
PayPal
about founding, 185–187, 291–292
acceptance of Bitcoin, xii, 261–262
Bitcoin support from, 129, 158–159, 184–185, 192, 349
buying/selling Bitcoin through, 38, 52–54, 110
ransom demands and criminal use, 347–348
restrictions by Argentina, 159–161
shutting down Mt. Gox account, 64
WikiLeaks blockade, 57
Paysius, 174
PC World, 57
People's Bank of China, 273–275
Pidgin (chat service), 246
Pirate Party, 35, 333
Ponzi scheme, Bitcoin as, 220
pornography, 72, 112, 117, 126, 234
Powell, Jesse, 94–96, 103, 105, 127–128, 139, 252, 315, 337
Pretty Good Privacy (PGP), 10, 13
proof-of-work, 18–19
P2P Foundation, 30, 323–324
public-key cryptography, 9–10, 141, 185–186, 238, 248, 281, 320, 330

Q coin (Chinese virtual currency), 257, 260–261, 268

ransom demands/payments, 82, 150, 347–348
Reeves, Ben, 237–239
reusable proofs of work (RPOWs), 18–19
Reuters, 211
Ribbit Capital. *See* Malka, Micky
Ripple, 187, 325
redandwhite (screen name), 225–226, 245
Russia, 54, 135–136, 197

Sacks, David, 192
Salmon, Felix, 210–211
SatoshiDice (gambling site), viii, 136, 193, 224, 338
Satoshi Ltd., 174
Satoshi Nakamoto
  creation and promotion of "e-cash," 5, 20–22
  disappearance/search for, xiv, 60–62, 80–81, 141
  participation in forums, 55–56, 58–59
  unearthing identity, 319–324, 339–340
Schumer, Charles, 84, 269
SecondMarket. *See* Barry Silbert
Shared Coin, 270
Shasky Calvery, Jennifer, 235, 266
Shrem, Charlie. *See also* BitInstant
  arrest by federal agents, 298–300
  background and founding of BitInstant, 128–130
  lack of management skill, 220–224
  marginalization in Bitcoin future, 331–334
  vacating Bitcoin Foundation seat, 345
Silbert, Barry, 143–144, 147–149, 217–218, 300, 303–304, 314, 325–326

Silicon Valley Bank, 203–204, 305–306
Silk Road
  additional resources, 368n
  BitInstant transactions, 129–130
  creation and business concept, 69–73
  as fringe group experiment, 335
  government investigation, 84–85, 121, 169–171, 213, 227–229, 298–300
  growth and success, 115–121, 137–138, 167–168
  growth in membership, 75–77, 82–84
  hacker penetration, 169, 225–226
  seizure by FBI, 245–253
Silk Road 2.0, 269–270
Sirius-M (screen name). *See* Malmi, Martti
Slashdot, 47–51, 53, 58
Snoop Dogg (rapper), 297
Snowden, Edward, 271
The Social Network (movie), 145
Songhurst, Charlie, 184, 292
Spain, 330
Spitzer, Elliot, 186
SpongeBob SquarePants stickers, 39, 69
Srinivasan, Balaji, 191–192, 294–295, 329
scout (screen name), 169, 246
silkroad (screen name), 73, 118. *See also* Ulbricht, Ross
Stephenson, Neal, 19, 252
Summer Olympics (Beijing, 2008), 145
SVBitcoin (email list), 204
Swap Variety Shop, 39
Szabo, Nick, 18–19, 338–341, 351

Taaki, Amir, 57–58
Tanona, Bill, 165, 180
Target Corporation, data breach, 288–289

taxes/taxation, 13, 126–127, 168, 219–220, 239–241, 287
Tea Party movement, xi–xii
TechCrunch, 214
Tencent (Chinese Internet company), 261, 284–285
Texas Bitcoin Association, 331–334
Thiel, Peter, 185–187, 192, 211, 291
Tibanne (cat), 66, 140, 200, 312
Tibanne Ltd., 68
Time (magazine), 79–80
Tor (software/network), 71–73, 120, 245, 369n
Transaction malleability, 309–310
Trickster (screen name). *See* Malmi, Martti
21e6 (mining company), 191–192, 294–295, 329
Two Bit Idiot (blogger), 315

Ukraine, 329–330
Ulbricht, Lyn, 331–332
Ulbricht, Ross. *See also* Silk Road
about creation of Silk Road, 69–73
arrest by federal agents, 246–251
fundraising for legal defense, 331–332
murder-for-hire accusations, 225–226, 332
plans to go off-the-grid, 226–229
Underground Brokers (renamed Silk Road), 70. *See also* Silk Road
U.S. Central Intelligence Agency (CIA), 78, 81, 86–87
U.S. Department of Homeland Security (DHS), 121, 247
U.S. Department of Justice (DOJ), 186, 234, 266–267
U.S. Department of the Treasury. *See* Financial Crimes Enforcement Unit [FinCen]
U.S. Drug Enforcement Agency (DEA), 298
U.S. Federal Bureau of Investigation

(FBI), 137–138, 227–228, 245, 247–251
U.S. Federal Deposit Insurance Corporation (FDIC), 114
U.S. Federal Reserve
about role as U.S. central bank, 17, 23
assessment of Bitcoin, 266–267, 289, 328
function of gold standard, 31
monetary policy, 80, 110–111
technology, adaptation to, 132–133
2008 big bank bailout, 32, 286
U.S. Government. *See* Government regulation/investigation
U.S. Internal Revenue Service (IRS), 248. *See also* Taxes/taxation
U.S. Marshals Service, 353
U.S. National Security Agency (NSA), 271, 342
U.S. Secret Service (USSS), 17, 266–267
U.S. Securities and Exchange Commission (SEC), ix, 224, 338

Variety Jones [vj] (screen name), 118–119, 228
Vaurum (Bitcoin company), 340–341
Vavilov, Val, 330
Ver, Roger. *See also* Memory Dealers
background and intro to Bitcoin, 77–80
as Bitcoin spokesman, 214
dealing with ransom demands, 348
investment in BitInstant, 128–131, 175
investment in Blockchain.info, 237–239, 252
meeting Erik Voorhees, 107–110
NYC Bitcoin Expo 2011, 103–105
promotion of Bitcoin, 127–128, 294

background and intro to Bitcoin
(*cont.*)
  reaction to Mt. Gox collapse, 311,
    314
  relocation to Japan, 125–126
  renouncing U.S. citizenship, 126,
    169, 234, 330, 338
  responding to Mt. Gox hack,
    92–96, 308
  size of Bitcoin holdings, 287
  2013 Argentina, Bitcoin meeting,
    277
Vessenes, Peter, 138, 144, 200, 213,
  233
Virtual money. *See* Digital currency
Voorhees, Erik
  introduction to Bitcoin, 107–110
  early vision/prediction about
    value, vii–ix, xi–xiii
  gathering for Lake Tahoe poker
    game, vii–viii
  joining BitInstant, 130–132,
    135–137
  sale of SatoshiDice, viii, xv, 224
van Der Laan, Wladimir, 348

Wagner, Bruce, 102–104, 128
Walker, Paul, 325–326
Washington Post, 267
Wells Fargo Bank, 202, 219,
  272–273, 287–288, 302
WikiLeaks, xi, 56–58, 66–67, 80
Wikipedia, 4, 45

Williams, Tom (possible
  pseudonym), 98
Wilson, Fred, 154, 182, 212,
  300–301, 305
Winklevoss Capital, 149
Winklevoss, Tyler and Cameron
  backing BitInstant startup, ix,
    144–149, 173–177, 201–202,
    220–223, 297
  buying/selling Bitcoin, 180, 196,
    250–251
  investment in Bitcoin, 211–215
  loan to Mt. Gox, 205
  Mt. Gox collapse, 312–314
  regulatory filing for COIN, 353
  testifying at government hearing,
    300–302
Woo, David, 272
Woodside Bakery and Cafe,
  291–293
World War II, 31
Wuille, Pieter, 348

Xapo, 281–282, 292–296, 305–
  306, 349–351, 353. *See also*
  Wences Casares

Yang Linke, 255–256, 258, 260
Yoda (computer chip), 329

Zimmerman, Phil, 10
Zuckerberg, Mark, 145, 176, 221,
  291

ALLEN LANE
an imprint of
PENGUIN BOOKS

Recently Published

Slavoj Žižek, *Against the Double Blackmail: Refugees, Terror, and Other Troubles with the Neighbours*

Lynsey Hanley, *Respectable: The Experience of Class*

Piers Brendon, *Edward VIII: The Uncrowned King*

Matthew Desmond, *Evicted: Poverty and Profit in the American City*

T.M. Devine, *Independence or Union: Scotland's Past and Scotland's Present*

Seamus Murphy, *The Republic*

Jerry Brotton, *This Orient Isle: Elizabethan England and the Islamic World*

Srinath Raghavan, *India's War: The Making of Modern South Asia, 1939-1945*

Clare Jackson, *Charles II: The Star King*

Nandan Nilekani and Viral Shah, *Rebooting India: Realizing a Billion Aspirations*

Sunil Khilnani, *Incarnations: India in 50 Lives*

Helen Pearson, *The Life Project: The Extraordinary Story of Our Ordinary Lives*

Ben Ratliff, *Every Song Ever: Twenty Ways to Listen to Music Now*

Richard Davenport-Hines, *Edward VII: The Cosmopolitan King*

Peter H. Wilson, *The Holy Roman Empire: A Thousand Years of Europe's History*

Todd Rose, *The End of Average: How to Succeed in a World that Values Sameness*

Frank Trentmann, *Empire of Things: How We Became a World of Consumers, from the Fifteenth Century to the Twenty-First*

Laura Ashe, *Richard II: A Brittle Glory*

John Donvan and Caren Zucker, *In a Different Key: The Story of Autism*

Jack Shenker, *The Egyptians: A Radical Story*

Tim Judah, *In Wartime: Stories from Ukraine*

Serhii Plokhy, *The Gates of Europe: A History of Ukraine*

Robin Lane Fox, *Augustine: Conversions and Confessions*

Peter Hennessy and James Jinks, *The Silent Deep: The Royal Navy Submarine Service Since 1945*

Sean McMeekin, *The Ottoman Endgame: War, Revolution and the Making of the Modern Middle East, 1908–1923*

Charles Moore, *Margaret Thatcher: The Authorized Biography, Volume Two: Everything She Wants*

Dominic Sandbrook, *The Great British Dream Factory: The Strange History of Our National Imagination*

Larissa MacFarquhar, *Strangers Drowning: Voyages to the Brink of Moral Extremity*

Niall Ferguson, *Kissinger: 1923-1968: The Idealist*